New Theories of Everything

JOHN D. BARROW

New Theories of Everything

THE QUEST FOR ULTIMATE EXPLANATION

'I am very interested in
the Universe—I am specialising
in the Universe and all
that surrounds it'

— PETER COOK

OXFORD
UNIVERSITY PRESS

OXFORD
UNIVERSITY PRESS

Great Clarendon Street, Oxford ox2 6DP

Oxford University Press is a department of the University of Oxford.
It furthers the University's objective of excellence in research, scholarship,
and education by publishing worldwide in

Oxford New York

Auckland Cape Town Dar es Salaam Hong Kong Karachi
Kuala Lumpur Madrid Melbourne Mexico City Nairobi
New Delhi Shanghai Taipei Toronto

With offices in

Argentina Austria Brazil Chile Czech Republic France Greece
Guatemala Hungary Italy Japan Poland Portugal Singapore
South Korea Switzerland Thailand Turkey Ukraine Vietnam

Oxford is a registered trade mark of Oxford University Press
in the UK and in certain other countries

© John D. Barrow 2007

Published in the United States
by Oxford University Press Inc., New York

British Library Cataloguing in Publication Data

Data available

Library of Congress Cataloging in Publication Data

Data available

Typeset by SPI Publisher Services, Pondicherry, India
Printed in Great Britain
on acid-free paper by
Clays Ltd., St Ives plc

ISBN 978–0–19–280721–2

1 3 5 7 9 10 8 6 4 2

TO ROGER

Who still believes there should always be
something rather than nothing

Preface to the Second Edition

I was pleased to take the opportunity, provided by the Press, to prepare a new edition of Theories of Everything. Interest in this subject has continued unabated since my first attempts to explain their scope and limitations, and to place them in a broader cultural context than that of mathematical physics. Many new possibilities have emerged in the pursuit of a final string theory, and cosmology has taken an unexpected path into a realm populated by many other possible universes. Both developments have undermined the naïve expectations of many, that a Theory of Everythng would uniquely and completely specify all the defining quantities of the Universe that make it a possible home for life. There is a long way to go before the physicists' Theory of Everything is formulated and decisively tested. In the meantime, I hope that this extended survey of the newest developments will help point readers in the right direction and illuminate the way.

John D. Barrow
Cambridge, February 2007

Preface to the First Edition

'Everything' is a big subject. Yet modern scientists believe they have stumbled upon a key which unlocks the mathematical secret at the heart of the Universe: a discovery that points them towards a monumental 'Theory of Everything' which will unite all the laws of Nature into a single statement that reveals the inevitability of everything that was, is, and is to come in the physical world. Such dreams are not new; Einstein wasted the latter part of his life in a fruitless and isolated quest for just this Theory of Everything. But today such schemes are not to be found only on the desks of a few maverick thinkers and unconstrained speculators; they have entered the mainstream of theoretical physics and are worked upon by a growing population of the world's brightest young thinkers. This turn of events raises many deep questions. Can their quest really succeed? Can our understanding of the logic underlying physical reality be completed? Do we forsee a day when fundamental physics will be complete, leaving only the complex details latent within those laws to be unravelled? Is this truly the new frontier of abstract thought?

This book is an attempt to describe what the challenge facing Theories of Everything really is; to pick out those aspects of things which must be understood before we can have any right to claim that we understand them. We shall try to show that while Theories of Everything, as currently conceived, may well prove necessary if we are to understand the Universe around and within us, they are far from sufficient. We shall introduce the reader to those extra ingredients which we need to complete our understanding of what is, and in so doing we aim to display many new ideas and speculations which transcend traditional thinking about the scope and structure of scientific inquiry.

Numerous people have helped this book come to completion. The Senatus of the University of Glasgow invited the author to deliver a series of Gifford Lectures at the University of Glasgow in January of 1988 and this book elaborates upon the content of some of those lectures. I am particularly indebted to Neil Spurway for his gracious help with everything associated with those lectures. For advertent or inadvertent comments and discussions which have helped in the writing of this book I am grateful to David Bailin, Margaret Boden, Danko Bosanac, Gregory Chaitin, Paul Davies, Bernard d'Espagnat,

Jeffrey Friedman, Michael Green, Chris Isham, John Manger, Bill McCrea, Leon Mestel, John Polkinghorne, Aaron Sloman, John Maynard Smith, Neil Spurway, Euan Squires, René Thom, Frank Tipler, John Wheeler, Denys Wilkinson, Peter Williams, and Tom Willmore.

Writing a book can be a miserable business, not only for the author, but for all those in his immediate orbit. The most perceptive reflection upon this situation was one made by the late Sir Peter Medawar. It applies not only to the activities of authors, but to obsessives of many sorts: '... it is a proceeding that makes one rather inhuman, selfishly guarding every second of one's time and becoming inattentive about personal relationships; one soon formed the opinion that anyone who used three words where two would have done was a bore of insufferable prolixity whose company must at all times be shunned. A danger sign that fellow-obsessionals will at once recognize is the tendency to regard the happiest moments of your life as those that occur when someone who has an appointment to see you is prevented from coming.' Because of the danger of such distortions, family members require special thanks for their patience and forbearance in the face of frequent neglect. Elizabeth has supplied her constant support in innumerable ways; without it this work would never have begun. Finally, our children, David, Roger, and Louise, have shown a keen and unnerving interest in the progress of the manuscript without which the book would undoubtedly have been finished in half the time.

J.D.B.
Brighton, September 1990

ACKNOWLEDGEMENTS

Figure 7.3 is reprinted by permission of the publishers from *Mind Children* by Hans Moravec, Cambridge, Mass.: Harvard University Press, Copyright © 1988 by the President and Fellows of Harvard College. Figs 7.2 and 7.4 are adapted from the same source. Figure 7.6 is copyright © R. V. Solé, reproduced by permission of the artist.

Contents

CHAPTER I

Ultimate explanation

I have yet to see any problem, however complicated, which when you looked at
it in the right way, did not become still more complicated.
— POUL ANDERSON

AN EIGHTFOLD WAY

It seemed to me a superlative thing—to know the explanation of everything, why it
comes to be, why it perishes, why it is.
— SOCRATES

How, when, and why did the Universe come into being? Such ultimate questions have been out of fashion for centuries. Scientists grew wary of them;
theologians and philosophers grew weary of them. But suddenly scientists are
asking such questions in all seriousness and theologians find their thinking
pre-empted and guided by the mathematical speculations of a new generation
of scientists. Ironically, few theologians have an adequate training in physics
to keep abreast of the details, and few physicists have a sufficient appreciation
of the wider questions to make a fruitful dialogue easy. The theologians think
they know the questions but cannot understand the answers. The physicists
think they know the answers but don't know the questions. An optimist might
thus regard a dialogue as a recipe for enlightenment, whilst the pessimist
might predict the likely outcome to be a state in which we find ourselves
knowing neither the questions nor the answers.

Modern physicists believe they have stumbled upon a key which leads to
the mathematical secret at the heart of the Universe—a discovery that points
towards a 'Theory of Everything', a single all-embracing picture of all the laws
of Nature from which the inevitability of all things seen must follow with

unimpeachable logic. With possession of this cosmic Rosetta Stone, we could read the book of Nature in all tenses: we could understand all that was, is, and is to come. Of such a prospect, there has always been speculation but never confidence. But is this confidence now misplaced? This is one of the questions that the reader will be in a position to answer after turning the final page of this book. It is our intention to spell out the different ingredients that must comprise any scientific understanding of the Universe in which we live. These we shall find to be more diverse and slippery than has fondly been imagined by the purveyors of Theories of Everything. Of course, we must be circumspect in our use of such a loaded term as 'Everything'. Does it really mean everything: the works of Shakespeare, the Taj Mahal, the Mona Lisa? No, it doesn't. And the way in which such particulars of the world fit into the general scheme of things we shall discuss at some length in the pages to come. It is a vital distinction that needs to be made in our approach to the study of Nature. For we might like to know if there are things which cannot be straitjacketed into the mathematically determined world of science. We shall see that there are, and we will attempt to explain how they may be distinguished from the codifiable and predictable ingredients of the scientific world that will populate any Theory of Everything.

Scanning the past millennia of human achievement reveals just how much has been achieved during the last three hundred years since Newton set in motion the effective mathematization of Nature. We have found that the world is curiously adapted to a simple mathematical description. It is enigma enough that the world is described by mathematics; but by *simple* mathematics, of the sort that a few years energetic study now produces familiarity with, this is a mystery within an enigma.

Several are the reactions to this state of affairs. We could regard the Newtonian revolution as the discovery of a master key which opens doors faster with constant use. And although the pace of discovery has quickened dramatically in recent times, it will none the less continue to do so indefinitely. Our present pace of discovery of truths about seemingly fundamental things does not necessarily indicate that we are about to converge upon the spot where all the treasure lies buried. The process of discovery could continue indefinitely either because the complexity of Nature is truly bottomless or because we have chosen a particular way of describing Nature which, while being as accurate as we desire, is none the less at best always but an asymptotic approximation that only an infinite number of refinements could make correspond exactly to reality. More pessimistically, our human frame and its eventful evolutionary past may place real limits upon the concepts that we can accommodate. Why should our cognitive processes have tuned themselves to such an extravagant quest as the understanding of the entire Universe? Is it not more likely that the Universe is, in Haldane's words, 'queerer than we can ever know'? Whatever our speculations about our own position in the history of scientific discovery, we surely regard with a Copernican suspicion any idea that our human mental powers should be adequate to handle an understanding of Nature at its ultimate level. Why should it be *us*? None of the sophisticated ideas involved appear to offer any selective advantage to be exploited during the pre-conscious period of our evolution. Alternatively, we might take the optimistic view that our recent sucess is indicative of a golden age of discovery which will near completion during the early years of the next century. Thereafter, fundamental science will be more or less complete. True, there will be things left to discover, but they will be matters of detail, applications of known principles, polishing, elegant reformulation, or metaphysical rumination. Historians of science will look back at this and neighbouring centuries as the time when we discovered the laws of Nature.

We have been this way before. Perhaps there is a psychological desire to bring things to a successful completion as the end of each century approaches. Near the end of the last century, many also felt the work of science to be all but done. The Prussian patent office was closed down in the belief that there were no more inventions to be made. But some work carried out by a junior at another patent office in Berne changed all that and opened up all the vistas of twentieth-century physics.

Can we hope to give ultimate explanations of the Universe? Is there a *Theory of Everything* and what could it tell us? And just what would such a theory actually encompass? By their very nature, scientific investigations do not know their end from their beginning. We cannot tell how much of what at present we might be loath even to call science will need to be included in such an all-embracing picture of the world. Indeed, history teaches some interesting lessons in this respect. Today, physicists accept the atomistic viewpoint that material bodies are at root composed of identical elementary particles, as well supported by evidence. It is taught in every university in the world. Yet, this theory of physics began amongst the early Greeks as a philosophical, or even mystical, religion without any supporting observational evidence whatsoever. Thousands of years would pass before we even had the means to gather this evidence. Atomism began life as a philosophical idea that would fail virtually every contemporary test of what should be regarded as 'scientific'; yet, eventually, it became the cornerstone of physical science. One suspects that there are ideas of a similar groundless status by today's standards that will in the future take their place within the accepted 'scientific' picture of reality.

In the chapters ahead, we shall take a look at this quest for ultimate explanation and inquire a little into its ancient and modern precedents. We shall stress, unlike many other commentators, that, while knowledge of such a Theory of Everything, if it exists, is necessary in order to understand the physical universe we see about us, it is far from sufficient to achieve that goal. Other essential ingredients are required. Without them, our knowledge will always remain incomplete and partial, and our quest for ultimate explanation will remain unfulfilled. We shall see how our understanding of the Universe is influenced by eight essential ingredients:

- laws of Nature,
- initial conditions,
- the identity of forces and particles,
- constants of Nature,
- broken symmetries,
- organizing principles,
- selection biases, and
- categories of thought.

As our story develops, we shall enlarge upon the nature and contribution of these ingredients to the search for ultimate explanation. It is the author's naïve hope that some of the ideas that we shall encounter along the way may be of wider interest than merely as support for a cautious attitude towards the likely scope of any Theory of Everything. But before we begin to follow this eightfold way, let us begin at the beginning and look back at some of the first Theories of Everything and how their motivations have matured into those of the twentieth-century enquirers into the nature of things.

MYTHS

When I was a child, I spake as a child, I understood as a child, I thought as a child: but when I became a man, I put away childish things.
— ST PAUL

If you browse through the ancient mythological accounts of the origin of the world and the situation of its inhabitants, the overwhelming impression one obtains is of having wandered into a Theory of Everything. All around there is completeness, confidence, and certainty. There is a place for everything, and everything is in its proper place. Nothing happens by chance. There are neither gaps nor uncertainties. No room for progress; no room for doubt. All things are interwoven into a tapestry of meaning pulled taut by the cords of certainty. Surely these were the first Theories of Everything.

The term 'myth' has taken upon itself a meaning in everyday English usage that betrays its real content. It is a much maligned word. To call something 'a myth', to label a politician's assurances as 'mythical', is now just the journalese for saying these things are false or unreliable. Alternatively, we may simply bundle up myths with legends, fairy stories, and all manner of other fantastic or imaginative literature. But to do so is to miss a layer of meaning that is crucial for our enquiry. A myth is a story imbued with a meaning. The message it contains transcends the naïve medium of the story and allows the hearer to understand why things are as they are. By studying the myths of a particular culture, we do not learn anything terribly interesting about the origin of the Universe or of mankind in the way that their original hearers did; rather, we appreciate how they define the outer boundaries of the imagination of their authors. They reveal what things they have thought about, how far they have followed them, those things they see as important enough to merit explanation, and the extent to which they regard the world as a unity. Once we start asking what the details of these myths mean we have removed ourselves from the mindset of the original hearers. It is like asking the meaning of *Little Red Riding Hood*. No nursery child would dream of asking such a question: if they did so, they would cease to be a child. Like fairy tales, myths are meaningful at many unconscious levels. Too precise an analysis of their message and meaning would remove this multiplicity of layers and reduce the number of hearers who could be influenced by its messages. Myths do not arise from data or as solutions to practical problems. They emerge as antidotes for mankind's psychological suspicion of smallness and insignificance in the face of things he cannot understand.

Our modern attempts to explain everything within some all-encompassing scientific picture differ in certain subtle respects when compared with ancient speculative explanations. For the ancients, it was breadth alone that was the hallmark of success for their Theories of Everything. For us, it is breadth *and* depth that count. If we claim to explain everything that is found in the world by a system of thought which proposes that the whole Universe came into being one hundred years ago with all its complex components ready-made, but bearing all the features of having already existed for millennia, then we do indeed attain a breadth of 'explanation' but our explanation possesses no depth whatsoever. We can extract no more from our theory save what we put into it. A similar theory to the one just proposed was actually considered in the nineteenth century by Philip Gosse in an attempt to reconcile the conflict over fossil evidence for the Earth's great antiquity and widespread public belief in special creation having occurred only a few thousand years ago. Gosse proposed that the rocks appeared with the pre-aged fossils already present, bearing (false) witness to past generations of evolution. A deep theory, by contrast, is one which is able to provide explanations for a wide range of things with a minimal contribution being made to the conclusion by the number of input assumptions. The depth of a particular consequence could be characterized by the effort expended in performing the shortest chain of logical reasoning from the assumptions to the conclusion: the amount of waste heat that a computer would have to generate in the process of computing the answer from scratch.

The weakness of mythological Theories of Everything played a key role in their structure and evolution. If one has a weak explanation, then it lacks real explanatory power. As a result, each fresh fact that is discovered requires a new ingredient in order to weave it into the pre-existing tapestry. We see this displayed most clearly by the proliferation of deities in most ancient cultures. Each time a short chain of explanations ('Why is it raining?'—'Because the rain-god is crying') ends, it tends to end at a deity. In any attempt at ultimate explanation—whether it be mythological or mathematical—there are psychologically acceptable bottom lines. In most mythological stories, the entry of an overseeing deity marks an acceptable end to the backtrack of 'why' questions. The more arbitrary and disparate one's explanations for the events of Nature, so the more deities one will tend to invent.

At first, myths must have been simple and focused upon a single question. With the passage of time, they became intricate and unwieldy, bound only by the laws of poetic form. A new fantasy, a new god: one by one they can be added to the patchwork. There was no sense of the need for economy in the multiplication of arbitrary causes and explanations. All that mattered was that they fitted together in some plausible way. Today such patterns of explanation are not acceptable. Ultimate explanation no longer means only a story that encompasses everything.

An indiscriminate multiplication of deities creates other problems. It implies a conflict of legislation in the natural world. A picture of universal laws imposed upon the world by a Supreme Being will not easily emerge. Indeed, even when we look at the relatively sophisticated society of the Greek gods, we do not find the notion of an all-powerful cosmic lawgiver very evident. Events are decided by negotiation, deception, or argument, rather than by omnipotent decree. Creation proceeds by committee rather than by fiat. In the end, any appeal to such a moody collection of initial causes leads to the multiplication of *ad hoc* explanations, a spawning of unnecessary complexity that is going to require more of the same to keep it going in the future. There is no plausible route towards simplicity. By interlinking causes, by searching always for unity in the face of superficial diversity, modern scientific explanations prize depth above breadth. A deep and narrow theory can, and often does, graduate to become a deep and broad one. A broad and shallow theory never does.

It is not clear how we should regard the originators of the first mythological Theories of Everything. We tend to assume they were realists and hence at worst foolish, at best wrong, in their description of the world. But although most of their hearers undoubtedly did take such stories literally—indeed many people hold somewhat similar views today—there may well have been others who thought of them only as images of some unreachable truth, or cynics who saw them as useful fables or devices for maintaining the status quo.

Lest we relegate the myth-makers and their objectives to the miasmal mists of the past, we should remind ourselves of the way in which the desire for completeness of explanation continued down the centuries. The most striking example is that of the medievals with their bookish desire to codify and order everything that we know or ever could know of Heaven and Earth. Great systems like the *Summa* of Aquinas or Dante's *Divine Comedy* sought to unify all existing knowledge into a labyrinthine unity. Everything had a place; everything had a meaning. As C. S. Lewis observes, it was altogether a little too stifling:

> The human imagination has seldom had before it an object so sublimely ordered as the medieval cosmos. If it has an aesthetic fault, it is perhaps, for us who have known romanticism, a shade too ordered. For all its vast spaces it might in the end afflict us with a kind of claustrophobia. Is there nowhere any vagueness. No undiscovered by-ways? No twilight? Can we never really get out of doors?

And, just as primitive peoples found that unity and completeness led to a vast and unwieldy patchwork of uneasy alliances in order that everything could find a place, so the medievals' desire to harmonize all knowledge into a Theory of Everything became unmanageably complicated. Where the primitive mind met the challenge of completeness by imaginative invention and was then faced with the problem of fitting all these imaginings together, the medieval mind was fettered by its respect for existing books and authorities. It regarded the inherited written words of the ancient philosophers with the same ultimate authority that modern physicists attach to experimental evidence. But the sheer volume of these written authorities ensured that any unification of their philosophical thinking was a vast enterprise. The twentieth century is not immune to such desires either. We have only to look at the problems that had to be faced over the definition and meaning of mathematics near the turn of the century. The formalists wished to protect mathematics from paradox by making it a closed shop: they defined it to be the sum total of all the logical deductions made using all possible rules of inference from all possible starting assumptions. As we shall see in a later chapter, this attempt to trammel up all possible mathematical consequences proved impossible. The desire for completeness could not be realized even here, in the most formalized and controllable human empire of knowledge. This modern urge for completeness had developed hand-in-hand with the desire for a *unified* picture of the world. Where the ancients were content to create many minor deities, each of whom had a hand in explaining the origins of particular things, but might often be in conflict with one another, the legacy of the great monotheistic religions is the expectation of a single over-arching explanation for the Universe. The unity of the Universe is a deep-rooted expectation. A description of the Universe that was not unified in its mode of description, but fragmented into pieces, would invite our minds to look for a further principle which related them to a single source. Again, we notice that this motivation is essentially religious. There is no logical reason why the Universe should not contain surds or arbitrary elements that do not relate to the rest.

CREATION MYTHS

It is necessary to recognise that with respect to unity and coherence, mythical explanation carries one much further than scientific explanation. For science does not, as
its primary objective, seek a complete and definitive explanation of the Universe ... It
satisfies itself with partial and conditional responses. Whether they be magical, mythical or religious, the other systems of explanation include everything. They are applied
to all domains. They answer all questions. They account for the origin, for the present
and even for the evolution of the universe.
— FRANÇOIS JACOB

We are so familiar with myths and scientific explanations for everything around us that it is no easy task to place ourselves in the prehistoric mindset that existed before any such abstractions were commonplace. We might think that the alternatives available were simply to rely on reason or sight, or upon faith in some invisible personalities or spirits. But this is a false dichotomy. At such a primitive stage, it is very much an act of faith to seek any parallel between our thoughts and the way things are in the outside world. It is by no means obvious that the great impersonal forces of the natural world are amenable to discussion or explanation, far less to prediction. Indeed, so awesome and devastating are many of their effects that they might more persuasively appear to be an enemy or, worse still, the irrational forces of chaos and darkness.

It is with such scales lifted from our eyes that we should approach the ideas that evolved concerning the origins of the world that we find in the mythology and traditions of every culture. These stories are often exhibited as illustrating the prescience of a few ancients for some favourite modern idea like the creation of the Universe out of nothing or its infinite age; but there should be no serious intent behind such juxtaposition of ancient and modern. It is merely that distorted perspective on the past that finds it to be significant solely where it presages our present thinking.

Ancient cosmology was not scientific. Its *raison d'être* was neither to explain observations nor make predictions. Rather, it was to embroider a tapestry of meaning within which its authors could represent themselves, and with respect to which they could evaluate the status of the unknown and the mysterious. The organization of their local society could be justified and reinforced by making it commensurate with the story of the world's origin and form. The starkness of the contrast between their aims and ours is strikingly captured by Frances Yates:

> The basic difference between the attitude of the magician to the world and the attitude of the scientist towards the world is that the former wants to draw the world into himself, whilst the scientist does just the opposite, he externalizes and impersonalizes the world by a movement of will in an entirely opposite direction.

The primitive belief in order and in the sequence of cause and effect displayed by myths is consistent with the belief that it is necessary to have some reason for the existence of everything—a reason that pays due respect for the natural forces which hold life and death in their hands. If one's view of Nature involves a personification of natural forces, then this search for reason reduces to the attribution of blame. Such generalized assumptions by no means lead to a unique collection of ideas about how the Universe came into being. But if one scans all the known myths concerning the origins of the Universe, they reveal a surprisingly small collection of cosmogonical notions. We find rather rarely, and then somewhat ambiguously, a belief in creation of the world out of nothing, but we find also a belief in the restructuring of the world out of pre-existent chaos. Often it suffices for a story to explain the ordered world which we now see. The notion of explaining some pre-existent state from which the world was fashioned either is not called for or is recognized for the cul-de-sac that it will turn out to be. Occasionally, we find adherence to the notion of a cyclic pattern of history taking its cue from the diurnal and seasonal periodicities of the natural world or, more adventurously, to a world that had no beginning. Elsewhere, we encounter the picturesque idea that the world hatched from a 'cosmic egg' or appeared as the progeny of the embrace of two world-parents. In the same vein, we find a collection of traditions in which the world emerges from some primeval womb or is fished from the primordial waters of chaos by a heroic diver. Finally, there is a mythological pattern which embroiders the theme of some titanic figure engaged in a cataclysmic battle against opposing forces of chaos and darkness. Out of the heroic victory of light over darkness is born our own Cosmos.

All of these formulae for dealing with the existence of the world are happy to establish some initial cause beyond which explanations will not be sought. The cause is simple in that it is singular, whereas the world of experience is bewilderingly plural. These fantastic speculations differ from any modern scientific approach to the origin of things because they look to an ultimate purpose as part of the motivation or the initial mode of creation. Yet they share one aspect with modern attempts to understand the Universe. All begin as attempts to explain everything we see about us and find this quest leads inexorably back to the ultimate question: how did the Universe originate? Today, the real goal of the search for a Theory of Everything is not just to understand the structure of all the forms of matter that we find around us but to understand why there is any matter at all, to attempt to show that both the existence and the particular structure of the physical Universe can be understood, to discover whether, in Einstein's words, 'God could have made the Universe in a different way; that is, whether the necessity of logical simplicity leaves any freedom at all'.

ALGORITHMIC COMPRESSIBILITY

Irrationality is the square root of all evil.
— DOUGLAS HOFSTADTER

The goal of science is to make sense of the diversity of Nature. It is not based upon observation alone. It employs observation to gather information about the world and to test predictions about how the world will react to new circumstances, but in between these two procedures lies the heart of the scientific process. This is nothing more than the transformation of lists of observational data into abbreviated form by the recognition of patterns. The recognition of such a pattern allows the information content of the observed sequence of events to be replaced by a shorthand formula which possesses the same, or almost the same, information content. As the scientific method has matured, so we have become aware of more sophisticated types of pattern, new forms of symmetry and new types of algorithm that can miraculously condense vast arrays of observational data into compact formulae. Newton discovered that all the information he could possibly record about the motion of bodies in the heavens or on Earth could be encapsulated in the simple rules that he called the 'three laws of motion' together with his law of gravitation.

We can extend this image of science in a manner that sharpens its focus. Suppose we are presented with any string of symbols. They do not have to be numbers but let us assume for the sake of illustration that they are. We say that the string is 'random' if there is no other representation of the string which is shorter than itself. But we will say that it is 'non-random' if there does exist such an abbreviated representation. So, for example, if we take the string of numbers 2, 4, 6, 8, ..., and so on *ad infinitum*, then we can represent it more succinctly by recognizing it to be just the list of positive even numbers. It is clearly non-random. A short computer program could instruct the machine to generate the entire infinite sequence.

In general, the shorter the possible representation of a string of numbers, the less random it is. If there is no abbreviated representation at all, then the string is random in the real sense that it contains no discernible order that can be exploited to code its information content more concisely. It has no representation short of a full listing of itself. Any string of symbols that can be given an abbreviated representation is called *algorithmically compressible*.

On this view, we recognize science to be the search for algorithmic compressions. We list sequences of observed data. We try to formulate algorithms that compactly represent the information content of those sequences. Then we test the correctness of our hypothetical abbreviations by using them to predict the next terms in the string. These predictions can then be compared with the future direction of the data sequence. Without the development of algorithmic compressions of data all science would be replaced by mindless stamp collecting—the indiscriminate accumulation of every available fact. Science is predicated upon the belief that the Universe is algorithmically compressible and the modern search for a Theory of Everything is the ultimate expression of that belief, a belief that there is an abbreviated representation of the logic behind the Universe's properties that can be written down in finite form by human beings.

This reflection on the compressibility of Nature also nudges us towards an understanding of why mathematics is so useful in practice. Our scientific theories always seemed to be described by mathematics, and physicists seem only interested in Theories of Everything that are couched in the language of mathematics. Is this telling us something profound about the nature of the Universe or the nature of mathematics? It is simplest to think of mathematics simply as the catalogue of all possible patterns. Some of those patterns are especially attractive and are used for decorative purposes, others are patterns in time or in chains of cause and effect. Some are described solely in abstract terms, while others are made manifest on paper or in stone. When viewed in this way, it is inevitable that the world is described by mathematics. We could not exist in a universe in which there was no pattern or order of any sort. Some order is inevitable for us, and the description of that order (and all the other sorts that we can imagine) is what we call mathematics. So, the fact that mathematics describes the world is not a mystery, but the exceptional utility of mathematics is. It could have been that the patterns behind the world were of exceptional complexity which allowed no algorithms to be developed which approximated them in simple ways. Such a universe would 'be' mathematical but we would not find mathematics terribly useful in practice. We could prove all sorts of 'existence' theorems about what structures exist but we would not be able to predict the future in detail using mathematics in the way that mission control at NASA does. Seeing it in this light, we recognize that the great mystery about mathematics and the world is that such simple mathematics is so far-reaching. Very simple patterns, described by mathematics that is easily within our grasp, allow us to explain and understand a huge part of the Universe and the happenings within it. This is another way of saying that the Universe is extremely compressible in the algorithmic sense. An awful lot of its observed complexity can be reduced to the presence of very simple patterns, described by short formulae and small equations. In many ways the search for a Theory of Everything is a manifestation of a faith that this compression goes all the way down to the bedrock of reality, that the ultimate patterns that give the Universe its shape and feel will also be 'simple' in the sense that we can understand them and discover them. It relies on the complexity of our minds, and the reach of our technologies, being sufficient to understand and find those ultimate patterns. All things being equal, the most likely state of affairs would be that our capabilities are vastly more or vastly less than those required for the task. A situation in which we are *just* able to understand the ultimate patterns behind the Universe using contemporary mathematics has a suspiciously un-Copernican element to it—why are *we* so closely matched in complexity to the Universe.

The human mind is the device that allows us to abbreviate the information content of reality in this way. The brain is the most effective algorithmic compressor of information that we have so far encountered in Nature. It reduces complex sequences of sense data to simple abbreviated forms which permit the existence of thought and memory. The natural limits that nature imposes upon the sensitivity of our eyes and ears prevents us from being overloaded with information about the world. They ensure that the brain receives a manageable amount of information when we look at a picture. If we could see everything down to sub-atomic scales then the information-processing capacity of our brains would need to be prohibitively large. The processing speed would need to be far larger than it now is in order for bodily responses to occur quickly enough to evade dangerous natural processes. This we shall have more to say about in the final chapter of our story, when we come to discuss the mathematical aspects of our mental processing.

This simple picture of the process of scientific enquiry as the search for algorithmic compressions is a compelling one, but it is also a naïve one in many ways. In the chapters to follow, we shall see why this is so and explore the eight ingredients which we have already highlighted as being necessary for our understanding of the physical world, to show what role each plays in the modern quest for an all-encompassing picture of the world. We shall start with the oldest notion: that of the laws of Nature.

CHAPTER 2

Laws

Search well another world; who studies this.
— HENRY VAUGHAN

THE LEGACY OF LAW

We are the music-makers
And we are the dreamers of dreams
Wandering by lone sea-breakers
And sitting by desolate streams;
World-losers and world-forsakers,
On whom the pale moon gleams:
Yet we are the movers and shakers
Of the world forever, it seems.
— ARTHUR O'SHAUGHNESSY

Many threads entwined to form our concept of a law of Nature. At first, primitive societies and groups were impressed primarily by the irregularities of Nature: mishap, plague, and pestilence. In time, emphasis refocused upon the regularities of the environment and the means by which they could be most fruitfully exploited for advantage. Sense began to emerge from the welter of disparate natural phenomena. The irregularities became exceptions to, rather than conceptions of, the natural state of the world. It emerged that some degree of organization might lurk behind the ordered facets of the world just as it lay behind the ordered results of mankind's interventions in Nature.

Social and religious views coloured early ideas about the organization of the world. There were many paradigms. For some, the world was a living organism growing and maturing towards some great purposeful culmination. All its constituents contained innate imperatives which moved them to trace out the ways predestined for them. They followed not the rules of some external diktat but the manifestations of their immanent properties. The meanings of things were to be found in their ends, not in their present or past states. For others, the world was a cosmic city, ordered by transcendent laws and rules imposed by a Supreme Being. Moreover, it was a walled city within which order was preserved for our benefit. Beyond its borders lay chaos and evil. In other cultures, quite different ideas held sway. No outside lawgiver was imagined. No outside lawgiver was necessary. Instead, all things seemed to work together in harmony to compose the common good by mutual consent and interaction. The order in the world was seen as that of the ant colony, wherein every individual plays its part to produce a coherent self-intereacting whole. It is a spontaneous response to the requirements of the system as a whole, not the inflexible result of eternal and unchangeable laws of Nature.

Different modern cultures have been variously influenced by their religious heritage in coming to a satisfying picture of natural laws. In the Judaeo-Christian West, the influence of the divine lawgiver has been paramount. The laws of Nature are the dictates of a transcendent God. They enshrine faith in the existence of an underlying order to things. They sanction the investigation of Nature as a secular activity. They outlaw Nature gods and the potential conflicts of polygamous legislation in the Universe. Farther East, in cultures like that of the early Chinese, the dominant picture was more liberal in style, with Nature operating holistically to produce a harmonious equilibrium in which every ingredient interacts with its fellows to produce a whole that is more than the sum of its parts.

It is not hard to see why the Eastern holistic perspective made scientific progress so difficult. It denies the intuition that one can study parts of the world in isolation from the rest—that one can *analyse* the world—and understand a part without knowing the whole. In modern terms, the Western perspective has regarded Nature as a linear phenomenon in which what happens at a given place and time is determined exclusively by what has occurred at nearby places immediately beforehand. The holistic view assumes nature to be intrinsically non-linear so that non-local influences predominate and interact with one another to form a complicated whole. It is not that the Eastern approach was misguided. It was simply premature. Only very recently, aided by versatile computer graphics, have scientists come to terms with the description of intrinsically complex non-linear systems. A successful study of natural laws needs to start with the simple linear problems if it is ever to graduate successfully to the holistic complexities created by non-linearity.

Having drawn with broad brush-strokes the inter-relationship between religious beliefs and the wider philosophy of nature that it engenders within a society it is important to inject a note of caution. It is common for apologists to press the argument further and claim that modern science has emerged because of, or even from, the West's Christian religious roots. There is undoubtedly some grain of truth in this claim, rightly interpreted; but its uncritical acceptance is as mistaken as the common notion that religion and science have always been at war like the forces of darkness and light. The monotheistic basis for the concept of universal laws of Nature contains an element of the truth because modern science is something that has developed to fruition after the early events which shape religious history. Moreover, many great scientists were overtly religious and brought to their scientific work an explicit religious justification and motivation. While these facts cannot be denied, it is a giant leap to infer from this summary of events that modern science is therefore a necessary consequence of our Christian past which would not otherwise have arisen. Here, the apologist is seeking to persuade that the practice of science or the concept of universal laws is a logical outcome of a certain range of religious beliefs rather than merely something that has been fostered by them. Religious scientists, like Boyle, Newton, or Maxwell, undoubtedly existed in profusion, but they inevitably stressed those aspects of their religion which accorded well with their scientific intuitions and activities. They were satisfied that their work was in tune with a Christian view of the world in an age when the public face of religion was a far greater factor in people's lives than it is today. There were always other strands of Christian doctrine, less obviously convivial to the pursuit of theoretical science, which the very same scientists would subconsciously downplay or simply ignore. Others, who found science distasteful, materialistic, or even blasphemous, could always be found amongst the ranks of the theologians and philosophers. The virtues necessary for the successful pursuit of science are neither specifically nor exclusively those engendered by our Judaeo-Christian heritage, nor, indeed, by any other. To believe that science has necessary rather than actual religious precursors is to subscribe to a deterministic theory of history with unique effects and causes. The real world is immeasurably more complicated: it is a skein of many strands, knotted and tangled, whose beginning is out of reach and whose end we cannot know.

THE QUEST FOR UNITY

Man shall not join what God has torn asunder.
— WOLFGANG PAULI

As we have become more demanding of our explanations and pictures of the Universe, so we have found the scale of what we must explain to be far greater in extent than our predecessors could ever have imagined. As complexity has grown, so has physics fragmented into specializations, which in turn have found themselves partitioned into manageable pieces. Each has enjoyed its own successes in building up mathematical theories of the different fundamental forces of Nature and has endowed us with effective descriptions of each of the different interactions between particles of matter and light. The most striking aspect of these theories, beyond that of their huge success, is that until only recently they have been distinct in form and content, each compartmentalized from the others as though bearing witness to some curious paranoia in Nature. This goes against the grain of our belief in the unity of Nature.

Only very rarely have ambitious scientists attempted to construct a theory of physics which would unite all the disparate and successful theories of the different forces of Nature into a single coherent framework from which all things could in principle be derived. One of the earliest with a distinctly modern perspective was Bernhard Riemann, the nineteenth-century creator of the systematic study of non-Euclidean geometries. He envisaged a 'total theory of physics' united by mathematics, and wrote to Richard Dedekind of his belief that

> one can set up a completely self-contained mathematical theory, which proceeds from the elementary laws that are valid for individual points to processes in the actually given continuously filled space, without distinguishing whether it is gravity, electricity, magnetism, or the equilibrium of heat that is being treated.

The most famous modern attempts to implement it were those of Eddington and Einstein. They failed for many reasons. In retrospect, we recognize that knowledge of the elementary-particle world was then so seriously incomplete that neither Eddington nor Einstein were in a position even to see what needed to be unified, let alone how to do it. However, the flame they first ignited has remained glowing faintly in the background, often overshadowed by the fireworks provided by the latest advances in the understanding of particular pieces of nature, until being fanned into prominence by the most recent attempts by theoretical physicists to illuminate our picture of the Universe. Whereas past unifiers were regarded as lone eccentrics by their colleagues, tolerated because of the brilliance of their other contributions to physics, the unifiers of today populate the mainstream of physics and continually add to their number the most gifted young students. This is what distinguishes the physics of the 1980s from any that has gone before.

The current breed of candidates for the title of a 'Theory of Everything' hope to provide an encapsulation of all the laws of nature into a simple and single representation. The fact that such a unification is even sought tells us something important about our expectations regarding the Universe. These we must have derived from an amalgam of our previous experience of the world and our inherited religious beliefs about its ultimate nature and significance. Our monotheistic traditions reinforce the assumption that the Universe is at root a unity, that it is not governed by different legislation in different places, neither the residue of some clash of the Titans wrestling to impose their arbitrary wills upon the nature of things, nor the compromise of some cosmic committee. Our Western religious tradition also endows us with the assumption that things are governed by a logic that exists independently of those things, that laws are externally imposed as though they were the decrees of a transcendent divine legislator. In other respects, our prejudices reflect a mixture of different traditions. Some feel the force of the Greek imperative that the structure of the Universe is a necessary and inflexible truth that could not be otherwise, while others inherit the feeling that the Universe is contingent. In this connection, it is interesting to recall the commentary supplied by Charles Babbage the eccentric nineteenth-century pioneer of computing devices who was much exercised by the concept of the laws of Nature. He was the first to liken the Universe to a computer whose program (as we would now call it) comprised the laws of Nature; but this image provided him more readily with the conception of a different program or one which might turn up irregularities and novelty very occasionally:

> The more man inquires into the laws which regulate the material universe, the more he is convinced that all its varied forms arise from the action of a few simple principles. These principles themselves converge, with accelerating force, towards some still more comprehensive law to which all matter seems to be submitted. Simple as that law may possibly be, it must be remembered that it is only one amongst an infinite number of simple laws: that each of these laws has consequences at least as extensive as the existing one, and therefore that the Creator who selected the present law must have foreseen the consequences of all other laws.

Our attraction to that quality which we have come to call 'beauty', and which we associate with the detection of innate unity and harmony in the face of superficial diversity, has led us to expect that the unity of the Universe should be expressed in certain particular ways. If we are physicists we might often hear talk of the 'beauty' or 'elegance' of particular ideas or theories to such an extent that, like Dirac,* we make aesthetic quality a guide or even a prerequisite for the formulation of correct mathematical theories of Nature.

The aesthetic imperative of Dirac strikes the life scientist as strange, the more so when he discovers how ineffective physicists, for all their mathematical powers, so often prove to be when they stray into his menagerie. For physicists are used to dealing with the pristine symmetries and fundamental laws of Nature. This habit conditions them to seek and expect symmetry and mathematical elegance everywhere they look. But the living world is not a marble palace. It is the higgledy-piggledy outcome of natural selection and the competition between many interacting factors. The outcome is often neither elegant nor symmetrical.

ROGER BOSCOVICH

Dear Reader, you have before you a Theory of Natural Philosophy deduced from a single law of Forces.
— ROGER BOSCOVICH

Our picture of the physical world has expanded so rapidly during this century that it requires some effort to put oneself in the shoes of the scientist of a past century. For Newton, there was no classification of the different forces of Nature. Radioactivity and nuclear forces were unknown; electricity and magnetism were different observed phenomena. Until Newton united them, the terrestrial and celestial influences of gravity were conceptually quite distinct. Newton simplified our apprehension of the world by explaining all gravitational phenomena within a simple scheme which attributed the observed effects to the action of a single attractive force acting between all massive bodies. Despite the success of this programme, and the other areas of thermodynamics and optics in which Newton was able to bring logical simplicity to a plethora of confusing observations, he knew that there were

* On being asked what he meant by the beauty of a mathematical theory of physics, Dirac replied that if the questioner was a mathematician then he did not need to be told, but were he not a mathematician then nothing would be able to convince him of it.

areas still shrouded in mystery. He speculated that there must exist other forces of Nature—'very strong attractions'—which hold material bodies together, but he could take that intuition no further.

One of the most remarkable and neglected figures in the history of modern European science was Roger Boscovich. A Dalmatian Jesuit, at once a poet and architectural advisor to Popes, cosmopolitan diplomat and man of affairs, socialite and theologian, confidant of governments and Fellow of the Royal Society, but most of all a mathematician and scientist, Boscovich was a passionate Newtonian who was the first to have a scientific vision of a Theory of Everything. His most famous work the *Theoria Philosophiae Naturalis*, was first published in Vienna in 1758. After several editions, it culminated in the enlarged and revised Venetian edition of 1763. Its influence was wide and deep, especially in Britain, where Faraday, Maxwell, and Kelvin would record their indebtedness to its inspiration.

Boscovich aimed to extend Newton's overall picture of Nature in several important ways. In particular, he sought to 'derive all observed physical phenomena from a single law'. In so doing, he introduced a number of new concepts which still form part of the intuition of scientists. He emphasized the atomistic notion that Nature was composed of identical elementary particles and then aimed to show that the existence in Nature of larger objects with finite sizes was a consequence of the way their elementary constituents interact one with another. The resulting structures were equilibrium states between opposing forces of attraction and repulsion. This was the first serious attempt to understand the existence of solid objects in Nature. He saw that Newton's inverse-square law of gravitation alone was insufficient to explain the existence of structures with particular sizes because it endowed gravity with no characteristic scale of length over which its effects were especially manifest. The inverse-square law singles out no particular scale of length as special and has an infinite range. To explain objects of particular sizes requires a balance between gravity and some other force.

Boscovich proposed a grand unified force law which included all known physical effects. This was his 'Theory', as he called it. It approached the inverse-square law of Newtonian gravitation at large distances as required by observations of the lunar motions. But on smaller length scales, it is alternately attractive and repulsive and so gives rise to equilibrium structures whose sizes are dictated by the characteristic length scales built into the force law. The 'Law of Forces' he proposed is shown in Figure 2.1. Boscovich lays great stress upon the fact that this law is not merely a 'haphazard' aggregate of forces but needs to be a 'single continuous curve', which, he argues, witnesses to

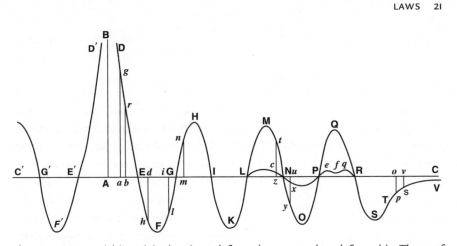

Figure 2.1 Boscovich's original universal force law, reproduced from his *Theory of Natural Philosophy*, first published in 1758. The variation of the force between two 'points of matter' as the distance between them changes is traced by the undulatory curve passing through the sequence of points DFHKMOQSTV. The distance between them is given along the abscissa AC; the strength of the force along the ordinate AB. The force is repulsive when this curve lies above the line AC and attractive when it lies below it. At very large distances (at and beyond V), it is attractive and approaches Newton's inverse-square law of force produced by gravity. The repulsive nature of the force as the separation of the two points tends to zero prevents all matter collapsing to zero size. Regarding this picture, Boscovich remarks: 'A Law of this kind will seem at first sight to be very complicated, and to be the result of combining together several different laws in a haphazard sort of way; but it can be of the simplest kind and not complicated in the slightest degree; it can be represented for instance by a single continuous curve . . . It is sufficient merely to glance at it.'

the unified all-encompassing nature of the theory. In addition to the pictorial representation of his force law illustrated here, Boscovich also introduced the idea of expressing his law as a convergent series of mathematical terms in powers of inverse distance, each smaller than its predecessor but the longer the sum is extended, the better becomes its approximation to the true force law.

There are many other innovations in Boscovich's detailed treatise, but we are interested here in drawing attention to just this one point: that he was the first to envisage, seek, and propose a unified mathematical theory of all the forces of Nature. His continuous force law was the first scientific Theory of Everything. Perhaps, in the eighteenth century, only a generalist like Boscovich, who successfully unified intellectual and administrative activities in every area of thought and practice would have the presumption that Nature herself was no less multicultural.

SYMMETRIES

But you see, I can believe a thing without understanding it. It's all a matter of training.
— (Lord Peter Wimsey in *Have His Carcase*)   DOROTHY SAYERS

For the early Greeks, the most perfect laws of Nature were its static harmonies. In the last two hundred years, the concept of a law of Nature has come to mean a set of rules which tell us how things change in space and in time. Thus, knowing the state of a system here and now, we seek a device for predicting its state at future times and in other places. But curiously, such laws of change can always be recast into completely equivalent statements which assert that something must not change: such unchanging quantities are known as *invariances*.

During the nineteenth century, mathematicians invested much time in classifying all the possible types of change and associated invariance that could exist, in both concrete and abstract terms. This classification gave rise to the branch of mathematics which we now call *group theory*. A 'group' is simply a collection of changes which possess three simple properties: there must be the possibility of no change, there must exist the possibility of undoing or reversing each change to restore its original state, and any two consecutive changes must give a result that could equally well be attained by another single change.

Each of the most basic physical laws that we know of corresponds to some invariance, which in turn is equivalent to a collection of changes which form a symmetry group. The symmetry group describes all the variations that can be formed from an initial seed pattern whilst still leaving some underlying theme unchanged. Thus, for example, the conservation of energy is equivalent to the invariance of the laws of motion with respect to translations backwards or forwards in time (that is, the result of an experiment should not depend on the time at which it was carried out, all other factors being identical); the conservation of linear momentum is equivalent to the invariance of the laws of motion with respect to the position of your laboratory in space, and the conservation of angular momentum to an invariance with respect to the directional orientation of your laboratory in space. Other conserved quantities in physics, which arise as the constants of integration of the laws of change, turn out to be equivalent to other less intuitive invariances of the laws of Nature. It is interesting to note that the conservation of energy was not used by Newton. Moreover, in the post-Newtonian discussions regarding the theological relevance of Newton's successful description of the world, the existence of conservation laws appears to have played some role in the growth of atheism amongst scientists. Some, like Newton himself, felt that there was need within the Newtonian dynamical model of the known universe (the solar system) for the sustaining and regulating hand of the Deity, but the subsequent discovery of conservation laws indicated that Nature possessed built-in sustaining principles which stopped the world from just ceasing to be. There were fewer roles for the Deity to play than had been believed. It was in this context that Laplace made his famous admission that '*nous n'avons pas besoin de cette hypothèse-là*' with regard to the sustaining role of the Deity in maintaining the motions within the solar system. Later the pendulum would swing back and the need to *violate* a conservation law of Nature in order to bring the Universe into being out of nothing persuaded many of the need for supernatural intervention. Moreover, the evident success of the concept of laws of Nature led to a reformulation of the Design Argument for the existence of God. We shall refrain from elaborating upon it here, but later, in Chapter 6, we shall return to highlight its special significance.

Even today there persists amongst many a feeling that the creation of the Universe out of nothing must violate some basic conservation law that stops one getting something for nothing. Nevertheless, there is actually no evidence that the Universe as a whole possesses a non-zero value of any such conserved quantity. The total mass-energy of all the constituents of a *finite* Universe appears to be always equal in magnitude but opposite in sign to the total gravitational potential energies of those particles. It could suddenly thus appear spontaneously without violating the conservation of mass-energy. Similarly, there is no evidence that the Universe possesses any overall net rotation or electric charge. It may well transpire that we discover some other conserved attribute that is non-zero for the Universe as a whole or obtain evidence that the Universe does indeed possess a non-zero electric charge or rotation. These ideas are based upon the supposition that the Universe is finite in size. Not only do we not know whether this is the case, but we cannot know because the finite speed of light ensures that we can only ever see a finite portion of the entire Universe. If the Universe were infinite in extent, then it is not known how one should associate conserved quantities with it and the question of whether it can appear out of 'nothing' without violating the conservation of charge, rotational momentum, and energy is a far subtler, unanswered question.

The fact that laws of change can be represented as invariances of the world under all possible changes that respect a particular innate pattern struck a resonant chord with physicists' expectations regarding the presence of symmetry and harmony in Nature. Symmetry has become the dominant theme in fundamental physics. Elementary-particle physics is singularly Platonic in this respect. Mathematicians of the past have catalogued all the distinct patterns of change that exist and have diligently encoded their essential ingredients into that branch of mathematics now known as group theory. By searching through its kaleidoscope of all possible patterns, the particle physicist can extract candidate symmetries to impose upon the world. The candidates need to pass some initial screening to ensure that they can accommodate all the necessary ingredients of the elementary-particle world and do not have some obvious consequence at variance with reality. The successfully vetted candidates then graduate to a more detailed mathematical outworking, which results in a gamut of predictions as to how particles should interact in a world governed by the imposed symmetry. Thus, a blind faith in symmetry provides an efficient recipe for generating candidate theories of elementary-particle interactions. No such machinery exists to generate candidate theories to explain the workings of less basic entities like economies or weather systems. The stronghold of symmetry is the unseen world of the smallest things.

Each of the four forces of Nature is accurately described by a theory that derives from the assumption of a particular invariance under all possible changes. The quest for unification proceeds by seeking to embed the separate patterns preserved by the several forces of Nature within a single 'Grand Unified' pattern into which the sub-patterns fit uniquely and completely. Such schemes are not easy to find and until recently carried with them unfortunate defects which came to light when the resulting pattern of invariance was used to compute observable quantities. Infinite answers were obtained which had to be dealt with in particular ways in order to produce sensible predictions.

So far, this flaw has been found to be absent only in a narrow class of unusual physical theories which have been proposed as the most complete laws of Nature by Michael Green, John Schwarz, and Edward Witten. These are known as 'superstring' theories. The prefix 'super' alludes to a powerful symmetry that they respect. This 'supersymmetry' has been proposed as a symmetry between otherwise distinct classes of elementary particles called fermions and bosons. In most situations, this amounts to a symmetry between matter and radiation. This idea was prevalent long before Green, Schwarz, and Witten. What they were able to do was wed it to the powerful concept of a 'string'.

Earlier theories of elementary particles had regarded the most elementary entities of Nature as point particles having no finite extent (they can be arbitrarily localized and they would offer no evidence of any internal structure when bombarded by other high-energy particles). They were described by quantum field theories in which the most basic elements are points of zero size. For the most part, they worked satisfactorily, but they were invariably beset by the disease of the infinities, and rather *ad hoc* mathematical remedies have had to be employed to suppress them. As time went on, they also became rather cumbersome: more and more quantum fields had to be introduced for each variety of elementary particle required to complete the picture. Strings tie things up more neatly. If the most elementary entities in Nature are regarded as strings (lines) rather than points, then all the unpleasant divergences in calculated quantities magically disappear for some very special universal symmetries. This reversal of fortunes arises because of the intrinsic differences in the way points and lines interact. In Figure 2.2 is displayed the schematic picture of a point particle and a string interaction in space and time. The particle interaction has obvious sharp corners that translate into mathematical infinities, whereas the smooth tube-like picture of the string interaction creates no such hiatuses. In effect, a certain collection of possible laws of Nature, and these only, may be finite and self-consistent.

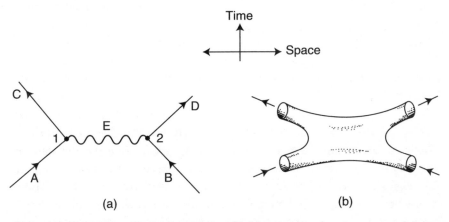

Figure 2.2 Diagrammatic representations of (a) interactions between two point particles A and B, mediated by the exchange of E, which results in the production of C and D; and (b) the interaction between two string loops which leads to two resultant strings. The diagrams represent the interactions in space and time with all the dimensions of space idealized to one for ease of presentation. As the point moves through space and time it traces out a line, whereas as the loops move through space and time they trace out tubes. The mathematical infinities associated with the point interaction arise because of the sharp corners at points 1 and 2 in (a). By contrast, the string interaction has no sharp corners at all and its smooth continuous character is a consequence of the absence of mathematical infinities in its calculation.

These fundamental strings possess a tension that varies with the energy of the environment in which they reside and this tension becomes large enough to shrink the loops of string to approximate points at the low energies we witness in the Universe today. But, in the extremities of the Big Bang, the essential stringiness of things should be manifest. It is theories of this sort that have aroused talk of finding a 'Theory of Everything'. If we had the correct version, it should in principle contain all the laws of radioactivity, gravity, electromagnetism, and nuclear physics. The enticing aspect of the string theories has been the unexpected discovery that the requirement of finiteness and consistency alone should prove to be so constraining.

Our attitude towards the laws of Nature and some ultimate codification of them into a possibly unique and self-consistently specified 'Theory of Everything' is a search for an ultimate symmetry of the world from within whose straightjacket there follow all the allowed causal laws of change governing the forces and particles of Nature. Our approach to such an apparent panacea must be tempered by an appreciation of how the laws of Nature—the Theory of Everything—might be related to the Universe.

INFINITIES—TO BE OR NOT TO BE?

Take it to the limit one more time.
—THE EAGLES

The boundless, timeless, and endless have attracted and confused human minds for thousands of years. From East to West, sophisticated cultures devised words to express the concept of infinity as well as arguments to include it or banish it from their models of the world. Aristotle first distinguished clearly between 'actual' and 'potential' infinities in the fourth century BC. The 'actual' variety, which involved infinite values of observable or measurable quantities here and now, were outlawed. On the other hand, he permitted the less threatening notion of a potentially infinite sequence, such as the unending list of positive whole numbers or an eternity of future time, where the infinity was neither achieved nor reached. It couldn't hurt you.

In many ways ancient attitudes towards actual infinities mirrored that towards the existence of a vacuum. For Aristotle, the two were intimately connected because in an empty space there would be no resistance to motion and bodies would eventually move with infinite speed. Therefore, no perfect vacuum could exist in the physical universe. Medieval scientists devised ingenious arguments to avoid either an infinity or a perfect vacuum ever occurring. A 'celestial agent' was imagined to act as a cosmic censor, ensuring that any opportunity to create a real vacuum or an infinity, even if permitted by the laws of mechanics, was never taken advantage of by Nature.

Much has happened to change our conceptions of the infinite since those path-breaking arguments emerged. But it still challenges theologians, philosophers, and scientists to understand it, cut it down to size, find out if it comes in different shapes and sizes, and to decide whether we want to outlaw it or welcome it with open arms into our descriptions of the Universe. Infinity is also very much a live issue. We have seen that physicists have spent the past twenty-five years searching for a Theory of Everything that unites all the known laws of Nature into a single mathematical statement. That search has been significantly guided by an attitude towards the existence of actual physical infinities.

In theories of particle physics, the appearance of an infinite answer to a question about the magnitude of a measurable quantity was always taken as a warning that you had made a wrong turn. For decades, the inevitable appearance of an infinity in the calculations was managed by a strange subtraction procedure that removed the divergent part from the calculation to leave only a finite residue to compare with observations. Although the results of this so-called 'renormalization' process gave spectacularly good agreement with experiments, there was always deep unease that this ugliness could not be part of Nature's economy. The true theory must be finite.

This all changed in 1984, when Green and Schwarz showed that a particular type of physics theory—a 'superstring' theory—could indeed be wholly finite. The enthusiasm with which the new theories were embraced by physicists was a consequence of their ingenious banishment of infinities, a problem that had plagued their predecessors.

The path towards superstring theories awaits experimental endorsement. But the energy with which they have been pursued reflects the philosophy of scientists who believe that the appearance of an actual infinity in a physical theory is a signal that it is being stretched beyond its domain of applicability. The usual response is to upgrade the theory until the infinities are smoothed into large, but finite, quantities. Engineers, for example, know this well. You can exorcise the appearance of infinities in simple models of rapid aerodynamic flows by simply including more realism in the description of the friction of the air. The crack of a whip is the sonic boom from the tip travelling faster than the speed of sound. A simple calculation that ignored the friction of air would say that this involved something changing infinitely quickly. But a more detailed modelling of the air-flow properties turns this infinity into a very rapid but *finite* change.

Despite the general adoption of this 'infinities-mean-you-must-try-harder' dictum in relation to physical theories, one area of science has been willing to take predictions of actual infinities more seriously. Cosmologists see there is room for a lot of infinities in the Universe. Many are of the 'potential' variety— the Universe might be infinite in size, face an infinite future lifetime, or contain an infinite number of atoms or stars. These are all potential infinities in Aristotle's sense. But there is one aspect of them that seems alarming to our common sense. While potential infinities pose no local threat to the fabric of reality, we do have to face up to Nietzsche's infinite replication paradox: if the universe is infinite in extent and exhaustively random, then any event that has a finite probability of occurring here and now (such as you reading this book) must be occurring infinitely often elsewhere at this very moment. Moreover, for every history we have pursued here, all possible alternatives are acted out, wrong choices made simultaneously with right choices.

This is a grave challenge to ethics and to the theology of almost every religion. Some find it so alarming that they regard it as a powerful argument for a finite universe. However, it should be remembered that the finiteness of the speed of light insulates us from contact with our doubles. We can only see and receive signals from a finite part of the universe. The distance that we would have to travel before we should expect to encounter a copy of ourselves is $10^{10^{28}}$ metres whereas the greatest distance that light has had time to reach us from is a mere 10^{27} metres. For all practical purposes, we experience a finite part of the Universe.

The challenge to cosmologists does not end there, though. They also have to worry about 'actual' infinities. For decades, cosmologists have been happy to live with the notion that the universe of space and time began expanding from an initial big bang 'singularity', where temperature, density and just about everything else, was infinite at some finite time in the past. Furthermore, when large stars exhaust their nuclear fuel and implode as a result of their own gravity, they appear doomed to reach a state of infinite density in finite time. But this is all neatly kept out of reach. Black holes are believed to be always shrouded by an 'event horizon'—a surface of no-return through which things can fall in but not pass out—so that we can neither see the infinite density at the black hole's centre, nor feel its effects, from the outside.

Roger Penrose, of Oxford University, believes that actual infinities do occur both at the start of the Universe and at the centre of black holes. He once proposed that the laws of Nature provide a form of cosmic censorship that ensures that such naked physical infinities are always enclosed by event horizons. This is reminiscent of the medievals' celestial agent invoked to avoid the creation of a perfect vacuum. It means that the effects of the physical infinity are confined to the inside of the black hole and cannot influence the outside world. The cosmological infinity at the beginning of the Universe is the one that influences us and, on this picture, could determine everything about the Universe that we see today. By contrast, cosmologists with a particle-physics perspective tend to see these black-hole and cosmological infinities merely as a signal that the theory has overextended itself and needs to be improved to exorcise these infinities. As a result, we find much interest in the prospect of universes that bounce back into expansion if run backwards in time towards their apparent beginning. Our presently expanding Universe is suspected by some cosmologists of having arisen from the rebound, at finite density and temperature, of a previously contracting phase in its history.

From the outside, we cannot see what is happening inside a black hole. But if we fell in, we would be facing an uncertain fate as we approached the centre. Is there a real physical infinity waiting there, or does energy slip away into another dimension of space, or simply disappear into nothing, or get soaked by exciting a never-ending sequence of vibrations of the superstrings at the core of all matter and energy? We just do not know. But again the issue of finite versus infinite is a crucial guiding principle. Do we treat the appearance of an infinity as a signal to update our theory, or do we treat it more seriously as an indication that new types of law govern infinite physical quantities, laws that could dictate how our Universe began and how matter meets its end under the relentless implosion of gravity.

Cosmologists have another strange potential infinity to contemplate: the possibility of an infinite future. Is the Universe on course to last forever? Its contemplation leads quickly to philosophy, for what does 'forever' mean? And to biology and computer science—or can life, in any form, continue forever? And to the social sciences—and what would it mean socially, personally, mentally, legally, materially, and psychologically for us to live forever? The last question, at least, is one to which we can all think of answers. In the long run, living forever might not prove as attractive as it seems at first. You might even welcome the televangelist who offers you the promise of finite life.

Mathematicians have also had to face up to the reality of infinity. Twice in the last 120 years, mathematics has faced civil war over the matter, leaving many a casualty and much bitterness. Some wished to outlaw actual infinities and redefine boundaries to forbid all treatment of them as real 'things'. Journals were compromised and mathematicians ostracized.

As we shall see in the next chapter, the nineteenth-century German mathematician Georg Cantor first showed how to make sense of the paradoxes of infinity. He elegantly defined infinite collections as those that could be put in one-to-one correspondence with subsets of themselves. This enabled him to go on and answer deeper questions: Can one infinity be bigger than another? Is there an ultimate infinity beyond which nothing bigger can be constructed or conceived, or do infinities go on forever? Cantor answered all these questions in precise ways, but did not live long enough to see the fruits of his genius form part of the acknowledged body of mathematics. He was sidelined and undermined by influential finitist opponents, and for long periods he turned instead to the study of history and theology and suffered bouts of depression before his death in 1918.

Remarkably, theologians were the first to seize on the importance of Cantor's work. They had long struggled to make sense of the infinities lurking within their doctrines. Is God alone infinite? Must he not be 'bigger' than other more mundane infinities? Many investigations of the infinite had been unpopular because they seemed to be challenging the belief that only God was infinite. Cantor's work changed all that. He revealed that there is a never-ending hierarchy of infinities, each unambiguously bigger than the last. This enabled us to distinguish between three different types of infinity: the mathematical, the physical, and the transcendental. Some thinkers accept them all, some accept only some, some accept none. The ancients, beginning with Zeno, were challenged by the paradoxes of infinities on many fronts. But what about philosophers today? What sort of problems do they worry about? There are live issues on the interface between science and philosophy that are concerned with whether it is possible to build an 'infinity machine' that can perform an infinite number of tasks in a finite time. Of course, this simple question needs some clarification: What exactly is meant by 'possible', 'tasks', 'number', 'infinite', 'finite' and, by no means least, by 'time'? Classical physics appears to impose few physical limits on the functioning of infinity machines because there is no limit to the speed at which signals can travel or switches can move. Newton's laws allow an infinity machine. This can be seen by exploiting a discovery about Newtonian dynamics made in 1971 by the US mathematician Jeff Xia. First take four particles of equal mass and arrange them in two binary pairs orbiting with equal but oppositely directed spins in two separate parallel planes. Now introduce a fifth much lighter particle that oscillates back and forth along a perpendicular line joining the mass centres of the two orbiting binary pairs. Xia showed that such a system of five particles will expand to infinite size in a *finite* time! How does this happen? The little oscillating particle runs back and forth between the binary pairs, each time creating an unstable meeting of three bodies. The lighter particle then gets kicked back, and the binary pair recoils outwards to conserve momentum. The lighter particle then travels across to the other binary and the same *ménage à trois* is repeated there. This continues without end, accelerating the binary pairs apart so strongly that they become infinitely separated while the lighter particle undergoes an infinite number of oscillations in the process.

Unfortunately (or perhaps fortunately), this behaviour is not possible when relativity is taken into account. No information can be transmitted faster than the speed of light and gravitational forces cannot become arbitrarily strong in Einstein's theory of motion and gravitation; nor can masses get arbitrarily close to each other and recoil—there is a limit to how close separation can get, after which an 'event horizon' surface encloses the particles to form a black hole. Their fate is then literally sealed—no such infinity machine could send information to the outside world. But this does not mean that all relativistic infinity machines are forbidden. Indeed, Einstein's relativity of time that is a requirement of all observers, no matter what their motion, opens up some interesting new possibilities for completing infinite tasks in finite time. Could it be that one observer could move fast enough to see an infinite number of computations occurring in a finite amount of their lifetime?

The famous motivating example of this sort is the so-called *twin paradox*. Two identical twins are given different future careers. Tweedlehome stays at home while Tweedleaway goes away on a space flight at a speed approaching that of light. When they are eventually reunited, relativity predicts that Tweedleaway will find Tweedlehome *to be much older*. The twins have experienced different careers in space and time because of the acceleration and deceleration that Tweedleaway underwent on his round trip. Time passes more slowly on the accelerated and decelerated trip. So can we ever send a computer on a journey so extreme that it could accomplish an *infinite* number of operations by the time it returns to its stay-at-home owner? Itamar Pitowsky, a philosopher of science at the Hebrew University of Jerusalem, argued that if Tweedleaway could accelerate his spaceship sufficiently strongly, then he could record a finite amount of the Universe's history on his own clock while his twin records an infinite amount of time. Does this, he wondered, permit the existence of a 'Platonist computer'—one that could carry out an infinite number of operations along some trajectory through space and time and print out answers that we could see back home. Alas, there is a problem— for the receiver to stay in contact with the computer, it also has to accelerate dramatically to maintain the flow of information. Eventually the gravitational forces become stupendous and it is always torn apart if it has any finite size.

Notwithstanding these problems a check-list of properties has been compiled for universes that can allow an infinite number of tasks to be completed in finite time, or 'supertasks' as they have become known. These are called

Malament–Hogarth (MH) universes after David Malament, a University of Chicago philosopher, and Mark Hogarth, a former Cambridge University research student, who in 1992 investigated the conditions under which super-tasks were theoretically possible. Supertasks open the fascinating prospect of finding or creating conditions under which an infinite number of things can be seen to be accomplished in a finite time. This has all sorts of consequences for computer science and mathematics because it would remove the distinction between computable and uncomputable operations.

It is something of a surprise that MH universes are possibilities but, unfortunately, they have properties that suggest they are not realistic unless we embrace some disturbing notions, such as the prospect of things happening without causes, and travel through time. The most serious by-product of being allowed to build an infinity machine is rather more alarming though. Observers who stray into bad parts of these universes will find that being able to perform an infinite number of computations in a finite time means that any amount of radiation, no matter how small, gets compressed to zero wavelength and amplified to infinite energy along the infinite computational trail. Thus any attempt to transmit the output from an infinite number of computations will zap the receivers and destroy them.

So far, these dire problems seem to rule out the practicality of engineering a relativistic infinity machine in such a way that we could safely receive and store the information. But the universes in which infinite tasks are possible in finite time include a type of space (called 'anti-de Sitter space') that plays a key role in the structure of the very superstring theories that looked so appealingly finite. Perhaps infinity still lurks in the wings ready to play a new and unexpected role in the drama of the Universe.

FROM STRINGS TO 'M'

'But do you really mean, sir,' said Peter, 'that there could be other worlds—all over the place, just around the corner—like that?' 'Nothing is more probable', said the Professor, taking off his spectacles and beginning to polish them, while he muttered to himself, 'I wonder what they *do* teach them at these schools.'
– C. S. LEWIS, *The Lion, The Witch, and The Wardrobe*

After the initial excitement that followed the proofs that string theories are finite, many years of detailed study followed with hundreds of young mathematicians and physicists flocking to join this research area at the world's leading physics departments. Progress was slow and difficult. It emerged that there were five varieties of string theory available to consider as a Theory of Everything, all finite and logically self-consistent, but all different. This was a little disconcerting. You wait nearly a century for a Theory of Everything then, suddenly, five come along all at once. They had exotic sounding names that described aspects of the mathematical patterns they contained— type I, type IIA, and type IIB superstring theories, SO(32) and E8 heterotic string theories, and eleven-dimensional supergravity. These theories are all unusual in that they have ten dimensions of space and time, with the exception of the last one, which has eleven. Although it is not demanded by the finiteness of the theory, it is generally assumed that just one of these ten or eleven dimensions is a 'time' and the others are spatial dimensions. Of course, we do not live in a ten- or eleven-dimensional space so in order to reconcile such a world with what we see it must be assumed that only three of the dimensions of space in these theories became large and the others remain 'trapped' with (so far) unobservably small sizes. It is remarkable that in order to achieve a finite theory we seem to need many more dimensions of space than those that we are aware of. This might be regarded as a prediction of the theory. It is a consequence of the amount of 'room' that is needed to accommodate the patterns governing the four known forces of Nature inside a single one without them being able to hive themselves off into sub-patterns that only talk to themselves rather than to everything else. Nobody knows why three dimensions (rather than one or four or eight, say) became large nor whether the number of large dimensions is something that arises at random (and so could be different— and may be different elsewhere in the Universe) or is an inevitable consequence of the laws of physics that could not be otherwise without destroying the logical self-consistency of the theory. One thing that we do know is that only in spaces with three large dimensions can things bind together to form structures like atoms, molecules, planets, and stars. No complexity and no life is possible except in spaces with three large dimensions. So, even if the number of large dimensions is different in different parts of the Universe, or separate universes are possible with different numbers of large dimensions, we would have to find ourselves living in one with *three* large dimensions no matter how improbable that might be, because we could exist in no other.

At first, it was hoped that one of these theories would turn out to be special and attention would then narrow in to reveal it to be the true Theory of Everything. Unfortunately, things were not so simple and progress was slow and unremarkable until Edward Witten, at Princeton, discovered that these different string theories are not really different. They are linked to one another by mathematical transformations that amount to exchanging large distances for small ones, and vice versa in a particular way. But this revealed that the five string theories were not the fundamental things that physicists had been searching for. Instead, they were each limiting situations of another deeper, but as yet unfound, Theory of Everything, which lives in eleven dimensions of space and time. That theory became known as 'M Theory', where M has been said to be an abbreviation for Mystery, Matrix, or Millennium, just as you like. We can think of M theory as the ball in Figure 2.3 and parts of its surface reveal the five string theories as limiting situations, cast like shadows upon it. The presence of the eleven-dimensional supergravity theory on the surface means that it might be that the hidden M theory is also eleven-dimensional but looks ten-dimensional at some places on its surface.

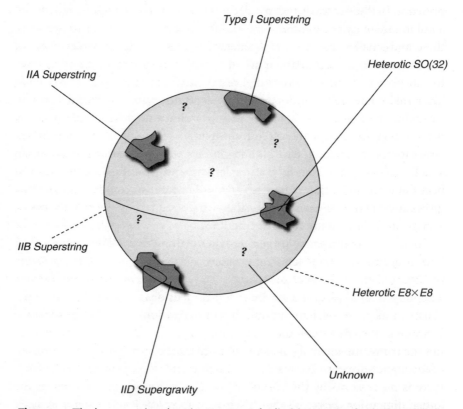

Figure 2.3 The known string theories appear to be limiting cases of a deeper underlying 'M' theory that has yet to be found. Each theory is described by a mathematical symmetry it displays. All of them exist in ten dimensions of space and time only, except for IID supergravity, which exists only in eleven dimensions.

These mathematical discoveries launched an intensive search for the under-lying M theory. But so far it has not been found. Other possibilities have emerged along the way, with the arguments of Lisa Randall and Raman Sun-drum that the three-dimensional space that we inhabit may be thought of as the surface of a higher-dimensional space in which the strong, weak, and electromagnetic forces act only in that three-dimensional surface while the force of gravity reaches out into all the other dimensions as well. This is why it is so much weaker than the other three forces of Nature in this picture.

Do these 'extra' dimensions of space really exist? This is a key question for all these new Theories of Everything. In most versions, the other dimensions are so small (10^{-33} cm) that no direct experiment will ever see them. But, in some variants, they can be much bigger. The interesting feature is that only the force of gravity will 'feel' these extra dimensions and be modified by their presence. In these cases the extra dimensions could be up to one hundredth of a millimetre in extent and they would alter the form of the law of gravity over these and smaller distances. Big changes in Newton's inverse-square law of gravitational attraction between masses would occur, changing to an inverse-fourth power of the separation between masses, for example. This sounds like a major change but unfortunately it is very difficult to test. Gravity is so weak that the form of the law of gravity is untested at these tiny distances. It is too difficult to isolate the gravitational forces from all the overwhelmingly larger forces of adhesion, friction, magnetism, and so forth that dominate on small scales—look at that fly walking on the ceiling, gravity is too weak to beat the forces of surface adhesion that hold his feet to the paintwork. This gives experimental physicists a wonderful challenge: test the form of the law of gravity on submillimetre scales.

There is another fascinating consequence of extra dimensions that we shall have more to say about in Chapter 5. It involves changes to the constants of Nature. One of the changes to our picture of the world that results from accepting that we live in a nine- or ten-dimensional space is that the true constants of physics live in that number of dimensions too. The ones that we measure in the laboratory and have been in the habit of calling constants are not the truly fundamental constants of Nature at all, they are merely shadows of the higher-dimensional reality cast on our three large dimensions. Indeed, there is no reason why they need be constant at all and we find that if the 'other' dimensions were to be slowly changing then we would see that because our three-dimensional 'constants' would change at the same rate as the change in the average size of the other dimensions. This means that observational searches for tiny variations in the traditional constants of Nature might reveal effects caused by wobbles or steady changes in other dimensions of space. Although we can't see them directly, in this case we can still see the effects of their existence in our own three-dimensional space.

A FLIGHT OF RATIONALISTIC FANCY

Why can't somebody give us a list of things that everybody thinks and nobody says, and another list of things that everybody says and nobody thinks.
— OLIVER WENDELL HOLMES

Let us examine some simple options that we can take with regard to the status of the laws of Nature. They provide a modern version of some ancient paradigms. Suppose, for simplicity, we restrict ourselves to three concepts: that of God (G) in the traditional omniscient and omnipotent sense; that of the Universe (U), taken to encompass the entire material world of space and time; and that of the laws of Nature (L), which prescribe its workings. The inter-relationships assumed between these three concepts rather succinctly encapsulate a number of different philosophies of Nature.

With regard to the pair U and L, we might choose one of five simple positions:

1. U is a subset of L;
2. L is a subset of U;
3. L is the same as U;
4. L is non-existent;
5. U is non-existent.

These are illustrated schematically in Figure 2.4.

The first option takes the laws of Nature to be something that transcends the physical Universe. The Universe is one of its particular manifestations. There may be others either in possibility or in actuality. It is important to notice that the recent direction of research in cosmology which has sought to provide a mathematical account of the creation of the Universe out of 'nothing' implicitly assumes the situation (1). It must assume that there pre-exist laws of Nature and other primitive notions like logic prior to the creation of the material Universe. If such a research programme were to be successful and come up with a self-consistent picture of the appearance of the physical Universe which made predictions repeatedly borne out by experiment, then the next research

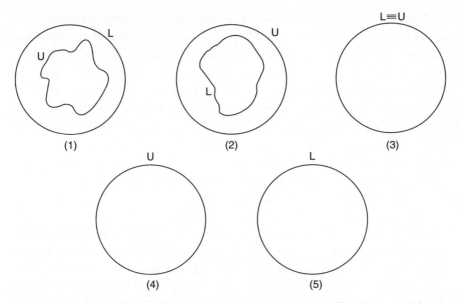

Figure 2.4 The possible relationships between the concepts of U, the material Universe, and L, the laws of Nature, investigated in the text.

programme would seek to understand why the laws of Nature, which allowed that, and no other, Universe to appear, do themselves exist and whether they could be different. This ultimate quest lies far in the future, but it is interesting to consider that if the Universe *as a whole* is described by a law of Nature like that enshrined in Einstein's general theory of relativity then there must exist a logical structure larger than the physical Universe. Certainly such an assumption is made implicitly in most cosmological studies. For one considers different possible mathematical models of the Universe each obeying the same laws of Nature but differing in their choice of starting states. Unfortunately, no observations could tell us whether any cosmological theory described by a set of mathematical equations really did describe the entire Universe, if only because we can only ever see a finite part of it.

If we subscribe to option (2), then we are nudged towards the view that the laws of Nature really possess some spatial or temporal dependence within the Universe. Elsewhere there may exist different laws or even none at all. There may exist islands of rationality within a possibly infinite universe. Since we know that the existence of observers like ourselves, and indeed observers considerably unlike ourselves, requires certain regularities to exist, we should not be surprised to find ourselves inhabiting one of the rational suburbs of such a chaotic universe. Attempts have been made to demonstrate that it is plausible to start the evolution of the Universe in a state which does not possess an exact adherence of things to some of its familiar laws and yet show that, as it expands, ages, and cools, behaviours at variance with what we have come to call laws of Nature will become rarer and rarer, so that in our low-energy world, fifteen billion years after the beginning of things, we observe an approximate adherence to certain patterns of behaviour that is so close to perfect that we assume it to be perfect. Taken to its logical conclusion, this philosophy would aim to show that all, or almost all, the observed laws of Nature are consequences of the late epoch of cosmic history at which we have come upon the scene. Back in the earliest moments of the Big Bang the situation would be largely lawless and very, very different.

Another, more sceptical, interpretation of the second alternative is to regard the laws of Nature as an invention of human minds, which themselves have emerged from the stuff of the Universe by natural processes. In different parts of the Universe, the historical process that led to this would necessarily be different: the environmental pressures would demand different responses and distinct evolutionary pathways would no doubt be followed. On this view, the laws of Nature are a creation, either in whole or in part, of minds and will thus vary from galaxy to galaxy in line with the distribution of sentient beings in the Universe. This view, whilst common enough amongst philosophers, has little to commend itself to scientists, because it does not lead to any future research programme which might test it, falsify it, or enlarge upon its content. It is something of a speculative dead-end. All one can do is await contact with hypothetical extraterrestrials and compare their 'laws' with our 'laws'.

The perspective (3) equates the Universe and the laws of Nature in a spirit that goes back at least as far as St Augustine and Philo of Alexandria, who avoided the problem of deciding what God was doing before the creation of the world by pointing out that there was no 'before' because time was part of the created order. Such a deduction involves the perception that time is not just measured by natural phenomena like the swinging of a pendulum but may in some deep sense always be associated with physical events within the Universe rather than imposed upon it as a transcendental back-drop. This leads to the natural conclusion that the Universe is coeval with time itself. In Philo's submission,

> Time began either simultaneously with the world or after it. For since time is a measured space determined by the world's movement, and since movement could not be prior to the object moving, but must of necessity arise either after

it or simultaneously with it, it follows of necessity that time also is either coeval with or later born than the world.

A similar perspective had been forced upon modern cosmologists up until recent years. Before attempts to understand quantum cosmology began in earnest, one was faced with the conclusion that our Universe must have experienced a space-time singularity at some finite moment in the past. Before this singularity, the Universe did not exist; afterwards, it did. The mathematical description of space and time predicts that both concepts must cease to exist at this singularity. It is the boundary of the Universe. Conversely, we are forced to regard universes which possess a past singularity as having an origin out of literally nothing at some past moment. At that moment, the material Universe, the laws of Nature, and the very fabric of space and time must come into being together.

It is important to stress that, although Einstein's general theory of relativity predicts that there can exist such a singularity in our past, it provides no reason why such a creation out of nothing should occur. If one does not want to come to terms with such a stark beginning to things, then there are ways of avoiding the conclusion that there existed a past singularity. If gravity were ever to become a repulsive, rather than an attractive, force in the distant past (and this seems rather likely given our present understanding of how matter could behave at very high energies), then the Universe need not have experienced a singular beginning. We offer this merely as an illustration of the perspective (3). We might also point out that this alternative may accommodate wider possibilities because the Universe is expanding and changing in time. Does this mean that we should, on this view, expect the laws of Nature to possess a reciprocal time variation? In fact, it is not logically possible for all the laws of Nature to be changing. Either there are no laws at all or there are invariant laws. Any changing law can always be traced back to the invariance of some more basic quantity which governs the rule of change. The alternative, that there exists no invariant bedrock, would mean that there could exist no laws of Nature at all. This leads us to our next option.

The fourth possibility, that there are no laws of Nature, is an extreme one. It might be defended in two ways. On the one hand, those of a more philosophical persuasion might seek to persuade us that what we choose to call the laws of Nature may be nothing more than the mental categories that our brains are forced to adopt in order to make sense of our experience. For all we know, there may exist no deep reality governed by true laws of Nature. Alternatively, a more realist perspective might be to imagine, as some physicists have done, that, as the Universe expands and ages from a state of chaos created by the simultaneous presence of all possible orders, some of these forms of order become predominant, so that after billions of years they dominate affairs so effectively that they pass for preordained laws of Nature rather than merely the stubbornest of possibilities. This possibility we have already mooted above.

This situation also includes one in which there are many universes governed by a single set of underlying laws. Each universe is an outcome of the laws for which some things can fall out differently. Until fairly recently, this scenario was a philosophically possible one with little scientific basis. But investigations of the wider consequences of the theory of the inflationary universe have provided the first reasons to take it a little more seriously. There is now good observational evidence that supports the idea that our visible universe underwent a surge of accelerated expansion, which we call 'inflation', in its very early stages. Observations of the small temperature variations in the microwave radiation left over from the early stages of the Universe display the same characteristic pattern of statistical variations that are predicted to result if we live in the vastly inflated image of a tiny primordial fluctuation. But this same theory makes other predictions that are not amenable to observational test. It predicts that the little fluctuations that inflate in the early Universe should continue producing further inflation from tiny parts of themselves over and over again. The process that results is an eternal self-reproduction process that creates a universe which is very different from place to place and at different times. We find ourselves living in a local 'bubble' that, like any of the others, may have had a beginning and may have an end. But the whole 'multiverse' of bubbles need have no beginning and no end. Each of the bubbles can differ in many respects. In the simplest versions of these theories the differences are in age and density. In other versions the differences are more fundamental: some bubbles have different numbers of fundamental forces of Nature and different numbers of dimensions of space. In these cases, the different bubbles are like different universes with some different laws even though they are all part of the same universal space. In effect, each of the bubbles is an outcome of the underlying laws of Nature which endow each bubble with some common features but many different ones. Some of those differing features are things that we have long regarded as so fundamental—like the number of dimensions of space—that they must be programmed into the Universe irrevocably, now turn out to be things that can fall out differently as outcomes of the laws of gravity and particle physics. All this is possible inside one universe or, in other theories still, in metaphysically separate 'other' universes governed by other laws that are logically self-consistent modifications of the ones that we know already. In all these situations the Universe—whether it is unique or one of many—is an outcome of the laws of Nature and, in a metaphysical sense, contained within them.

The last of our choices, that there is no Universe, is a peculiar form of nihilism that no earnest philosopher has ever taken seriously. However, it is of interest because if the quantum cosmological models which seek to create the Universe out of nothing are considered then this view encapsulates their 'pre-initial' state. One cannot therefore argue that such a position is logically impossible or self-contradictory, since it is an admitted precursor to the present state within this cosmological description. It may be unstable in some peculiar sense, but it is hard to see why it should be impossible. To argue in this direction would seem to take us perilously close to resurrecting the infamous Ontological Argument of Anselm and others, that there can exist concepts like that of a Supreme Being whose very conception necessitates their existence. This seems particularly dubious when one tries to conceive of how there could exist some entity whose non-existence would imply a logical contradiction.

For others, there is a tension primarily between the concepts of God and the Universe, rather than between the laws of Nature and the Universe. Indeed the concept of a Supreme Being is in all cultures a more primitive and natural notion than that of laws of Nature. It could well be argued that no culture arrived at a robust concept of the latter without a preliminary concept of the former. Again, it is convenient to list the naïve possibilities as follows:

(i)   U is a subset of G;
(ii) G is a subset of U;
(iii) G is the same as U;
(iv) G is non-existent;
(v)   U is non-existent.

The first option, that the Universe is part of God, is called *panentheism* in the terminology adopted by the eighteenth-century German philosopher Krause. Theologians distinguish this view from simple theism by associating the latter with a view that God is wholly other than the Universe, both above it and becond it. The panentheist believes that God is in all things but not identical to them.

The situation (ii) would be consistent with the sceptical attitude that the notion of 'God' is a creation solely of the human mind and hence of the purely material processes that gave rise to it. Alternatively, if the Deity were of the non-traditional sort and in some way limited to the role of a Super-being within the Universe, this situation would be approximated. There are many science fiction stories that have explored this paternalistic Superbeing scenario. The semi-religious vision of advanced forms of intelligent extraterrestrial life discussed by many enthusiasts for the search for extraterrestrial radio signals might also fit into this category.

The third possible relationship is associated with the doctrine of pantheism which regards God and the natural Universe as one and the same thing. This is a common view to be found in many non-personal Eastern religions and also amongst agnostic scientists; it is also a view with which Einstein professed some sympathy. It is what he means when he talks of his God being that of Spinoza, the philosopher most associated with the pantheistic view.

Our last two possibilities are easily dealt with. The option (iv) is the position of the atheist, whilst (v) has already been discussed as possibility (5) above.

The third side of our triangle of relationships consists of the possible interrelation of the laws of Nature and a Deity:

(a) L is a subset of G;
(b) G is a subset of L;
(c) G is the same as L;
(d) L is non-existent;
(e) G is non-existent.

The first case is in line with a Judaeo-Christian tradition that views the laws of Nature as constraints that God imposes upon the Universe. This was, for instance the view of Newton, who consequently maintained that the laws of Nature could have been different and could be suspended arbitrarily according to divine fiat.

The second possibility is somewhat akin to the schools of Process Theology that propose an evolving Deity. In this case, God is constrained by some higher-order logic. Although this would be difficult to reconcile with many pictures of an omnipotent Deity, it is difficult to draw the line between this position and what is generally assumed implicitly even in these pictures, that God's actions are bound by certain constraints of logic and related to such concepts as 'good' and 'evil'. Alternatively, this option may be interpreted non-theistically as one in which the concept of God is an inevitable outworking of the laws of Nature in the minds of certain species of complex biocomputers like ourselves.

The third case, which equates the laws of Nature with God, is similar to the impersonal picture of God adopted by some pantheists. But it also resembles the view of the Deists which emerged as a lowest common denominator in response to the labyrinth that seventeenth-century theologians of all creeds found themselves within. It reduced the number of attributes which the Deity was expected to display in the Universe and reduced Him to the role of initial cause and sustainer of the laws of Nature who thereafter maintained all things in harmonious development. The cases (d) and (e) we have already met as (4) and (iv) above.

GOODBYE TO ALL THAT

Every dogma must have its day.
— H. G. WELLS

Our exploration of the laws of Nature has been rather cursory.* The reason is deliberate: to most minds the issue of a Theory of Everything is about nothing more than the laws of Nature. It is a quest for the most basic and most comprehensive versions of those universal laws. From these, it is assumed that everything one might want to know or explain regarding the nature of the observed universe would follow with a little work. In the chapters to follow, we hope to undermine this dogma and reveal what other aspects of the physical world, distinct from the traditional image of laws of Nature, are needed to understand its overall structure. To come to terms with them will require either additional facets of the Universe to be uncovered or the concept of a law of Nature to be considerably deepened and widened to unify it to other concepts that are at present logically disjoint from it.

From Boscovich to superstrings, the searchers for a unified Theory of Everything have focused upon finding the all-encompassing laws of Nature to the exclusion of all else. At root, this prejudice has grown from an implicit subservience to the Platonic emphasis upon timeless universals as more important in the nature of things than the world of particulars that we observe and experience. In the chapters to follow, we shall examine the challenges to this view that are offered by our latest ideas about the physical world. The first is almost familiar. Since science pays homage to the gods of change, it needs to know how things began if it is to know anything at all.

* A more extensive study of this subject can be found in the author's earlier volume *The World within the World*.

CHAPTER 3

Initial conditions

Once upon a time and a very good time it was.
— JAMES JOYCE

AT THE EDGE OF THINGS

Science is a differential equation. Religion is a boundary condition.
— ALAN TURING

Laws of Nature tell us how things change. Yet behind them we believe there to lurk invariances that straitjacket reality. Nature can do whatever she pleases so long as these charmed quantities stay the same throughout the change. The Theory of Everything seeks to provide us with the ultimate directory of all possible changes. The guiding principle in the search for this all-controlling formula is that it must be a single law, not a collection of different pieces. The logical unity of the Universe demands a single invariance that remains unchanged in the face of all the complexity and transience we see about us from the smallest sub-atomic scales to the farthest reaches of outer space. Identifying this over-arching symmetry, if it does indeed exist and is manifest in a form that is intelligible to us, may be the nearest thing we could get to discovering the 'secret of the Universe'.

Yet this is still not enough. Even if we knew the rules which govern how all things change, then we can only understand the present structure of things if we know how they began. This is a legacy of our belief in the rule of cause and effect in the Universe and our representation of laws of Nature as differential equations or algorithms in which output is determined uniquely by input. Differential equations are mathematical 'machines' which allow us to predict the future from the present. Equally, they enable us to use the present to reconstruct the past.

AXIOMS

Set theory can be viewed as a form of exact theology.
— RUDY RUCKER

In mathematics, the role of initial conditions is played by axioms. These are the initial postulates that are made before we start employing any deductive reasoning. The classic example of an axiomatic system is that of plane geometry formulated by Euclid in about 300 BC. It forms the model of all rigorous mathematical schemes. The axioms are initial assumptions which are taken as self-evidently true. From them, logical deductions can proceed under stipulated rules of reasoning. These rules of logical reasoning are analogous to the scientists' laws of Nature, whilst the axioms play the role of initial conditions.

We are not free to pick any axioms we might care to choose. They must be logically consistent. But there is no limit to their number although the number of axioms that we introduce will determine the size and richness of the logical deductions that can follow from them. Whereas Euclid and most other pre-nineteenth-century mathematicians knew that logical consistency was essential in any choice of axioms, they were also strongly biased towards picking axioms which mirrored the way the world was observed to work. Thus Euclid's axioms—for example, that parallel straight lines never meet, or that there is only one straight line joining any two points on a flat surface—are the self-evident fruits of one's experience of drawing lines on a flat surface. Later mathematicians did not feel so encumbered and have required only consistency from their lists of axioms. They need have no correspondence with anything we can see or abstract from experience. It remains to be seen whether the initial conditions appropriate to the deepest physical problems, like the cosmological problem which we shall discuss below, will have specifications which are directly related to visualizable physical things, or whether they will be abstract mathematical or logical notions that enforce only self-consistency. Even if the latter situation prevails, it may transpire that the requirement of self-consistency in a system as self-evidently complex as the physical universe is adequate to fix those initial conditions uniquely and completely.

Another important lesson we have learnt from the mathematicians' approach to axiomatic systems is that one can quantify the amount of information that is contained in a collection of axioms. None of the possible deductions that can be proved from these axioms using the allowed rules of reasoning can possess more information than was contained in the axioms. In essence, this is the reason for the famous limits to the power of logical deduction expressed by Gödel's incompleteness theorem. The axioms of ordinary arithmetic (and any axiomatic system rich enough to contain the whole of arithmetic) contain less information than some arithmetical statements and hence those axioms and their associated rules of reasoning cannot determine whether these statements are true or false. Note, however, that an axiomatic system which is not as large as the whole of arithmetic does not suffer from Gödel's incompleteness. For example, the so-called Presburger arithmetic, which consists of the operation of addition upon zero and the positive whole numbers (but not subtraction) has the property that all its statements are decidable. Its reduced set of axioms contain sufficient information to ascertain the truth or falsity of all the statements that can be framed using its vocabulary.

In the first chapter, we introduced the notion of algorithmic compressibility as a criterion for determining the degree of randomness of mathematical expressions. We can make use of this concept again here to sharpen our discussion. If presented with a particular sequence then we cannot prove it to be random, although we can prove it to be non-random simply by finding a compression. The minimum compression that is possible for a logical system corresponds to the axioms of the system. Thus we see why there can be no theorem of the system which possesses a larger information content than the axioms of the system.

Axioms are not therefore quite as straightforward as one might have hoped. It is often a rather subtle question to decide whether different proposed axioms are truly independent of each other. There is one classic case of this sort which enshrouds one of the most difficult unsolved problems in mathematics. It is called the *continuum hypothesis*. Prior to the work of Georg Cantor in the mid-nineteenth century, mathematicians had denied the existence of real infinities. Indeed, infinities were an 'abomination' in the words of one famous mathematician. Gauss's views on the matter are that he would

> protest against using infinite magnitude as something consummated; such a use is never admissible in mathematics. The infinite is only a *façon de parler*: one has in mind limits which certain ratios approach as closely as is desirable, while other ratios may increase indefinitely.

Here we see spelt out the notion that the infinite can never be an actuality; it is merely a shorthand for something that can be as large as one wishes. But Cantor turned the world upside-down by treating infinities like other mathematical quantities and creating an entire series of infinities of different sizes. The smallest was the set of natural numbers $\{1, 2, 3, 4, 5, \ldots\}$, which was labelled $\aleph_0$ (aleph-nought). Another infinite set is said to have the same size (or *cardinality*) as $\aleph_0$ if its members can be put into a direct one-to-one correspondence with the natural numbers; that is, if they can be systematically counted. For example, the infinite set of all the even numbers $\{2, 4, 6, 8, 10, \ldots\}$ can be counted in this way by the correspondences displayed by the sequence of arrows in Figure 3.1(a). The arrowed path in Figure 3.1(b) then shows how all the rational fractions laid out in an infinite array can be counted one by one without any being omitted. This shows there to exist a direct one to one correspondence between $\aleph_0$ and all the rational fractions through the sequence $\frac{1}{1}, \frac{2}{1}, \frac{1}{2}, \frac{1}{3}, \frac{2}{2}, \frac{3}{1}, \frac{4}{1}, \frac{3}{2}, \frac{2}{3}, \frac{1}{4}, \frac{1}{5}, \frac{2}{4}, \frac{3}{3}, \frac{4}{2}, \frac{5}{1}, \frac{6}{1}, \frac{5}{2}, \frac{4}{3}, \ldots$, and so on *ad infinitum*. Hence, in this precise sense, the rational fractions are an infinite set of the same size as the natural numbers. At first sight, this is a surprising result, since natural numbers are rather sparsely distributed whereas there seem to be rational fractions densely packed everywhere in-between them so a counting process ought to find many more fractions that integers. But this intuition focuses too much upon the *order* in which the numbers appear, whereas the one-to-one correspondence that we have set up does not need to follow the order in size with which the fractions occur in between the integers. A fraction

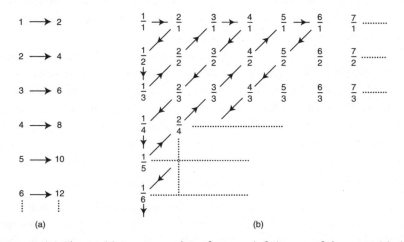

(a) (b)

Figure 3.1 (a) The positive even numbers form an infinite set of the same 'size' as all the positive integers because the two sets can be put into the direct one-on-one relationship shown here. This means that they can be systematically counted. (b) The set of all rational fractions can also be put in a one-on-one relationship with the positive integers and hence be systematically counted if they are listed in the pattern shown and then counted in the order marked by the sequence of arrows, *ad infinitum*.

is just specified by a pair of numbers and there are as many infinite pairs of numbers as there are numbers.

If we now try and count not just all the fractions but the decimals as well, then something qualitatively different happens because there are so many more decimals than fractions. The jump in size that marks the step from the natural numbers to the decimals is comparable to the step one would have to take from just the numbers zero and one to the larger ones. To take such a step, further information is required, because the only way that we can make 2 from 0 and 1 is to add two 1's together, but such a move requires us to be in possession of the concept of 'two' already.

Cantor showed that if we try to count the number of infinite decimals (the so-called 'real numbers'), then we fail. They are of a higher cardinality than the natural numbers and so cannot be placed in a one-to-one correspondence with them. This he showed by an ingenious and very powerful new form of argument. It involves the notion of a diagonal number. For illustration, suppose we have four numbers of four digits in length:

$$1234$$
$$5678$$
$$9012$$
$$3456$$

Then the diagonal number 1616 is not one of the four numbers listed. What Cantor showed was that if we make this array of numbers infinitely large then there is always a way of concocting a diagonal number that is not one of the infinite list of numbers lined up to make the array. Suppose we just look at the real numbers between zero and 1 (it does not make any difference to the basic argument if we add all the others as well) and suppose that we can count all the infinite decimals. This, Cantor showed, leads to a contradiction. Suppose we could write down all possible infinite decimals and align them one-to-one with the natural numbers. Let us suppose the list begins as follows:

1	0.234566789	...
2	0.575603737	...
3	0.463214516	...
4	0.846216388	...
5	0.562194632	...
6	0.466732271	...

and so on to infinity. Now take the diagonal number with decimal part composed of the highlighted digits:

$$0.273292 \quad ...$$

Next, alter each digit by adding one to it to get the new decimal

$$0.384303 \quad \ldots$$

Then this new number cannot appear anywhere on the original list because it differs by one digit from every single horizontal entry.* It must at least differ from the first entry in the first digit and from the second entry in the second digit and so on. So, contrary to our original supposition, the list could not have contained all the possible decimals. Hence, our original assumption that all the infinite decimals can be systematically counted was false. The real numbers possess a higher cardinality than the natural numbers and it is denoted by the symbol $\aleph_1$ ('aleph-one').

Cantor raised the intriguing question of whether there exist infinite sets which are intermediate in size between the natural numbers and the real numbers. Cantor thought that there could not be, but was unable to prove it. This is called the *continuum hypothesis*. Indeed, Cantor appears to have suffered a mental breakdown brought about by the intellectual effort he expended upon this question. The problem remains unsolved to this day. Nevertheless Kurt Gödel and his young American colleague Paul Cohen demonstrated some deep and unusual things about it. Gödel showed that if we merely treat the continuum hypothesis as an additional axiom and add it to the conventional axioms of set theory† then no logical contradiction can result. But then, in 1963, Cohen showed that the continuum hypothesis is independent of the

* As a technicality note that we have to remove any ambiguity about decimals that end with recurring 9s because $0.2399999999\ldots$, say, is the same as $0.240000000\ldots$. So, by cutting out those decimal expansions that end in a run of zeros we identify the rational number 24/100 by $0.23999\ldots$ and not by $0.240000\ldots$.

† The seven axioms of standard set theory which are intended to be sufficient for the deduction of all of mathematics (and hence for the mathematical representation of physics) are as follows. (1) *Extensionality*: two sets are equal if and only if they contain the same members. (2) *Subsets*: given a set S and some meaningful property, there exists a set containing members of S, and only those members of S which possess this property. (3) *Pairing*: given any two different sets, there exist another set that contains just the members of these two sets. (4) *Sum-set*: if there is a set S whose members are themselves sets, then there exists a set (called the sum set of S) whose members are just the members of the members of S. (5) *Infinity*: there exists at least one infinite set (i.e. the natural numbers 1, 2, 3, ...). (6) *Power set*: for any set S, there exists another set whose members are the subsets of S. (7) *Choice*: if S is a set of sets that is not empty and no two distinct members of S have an element in common, then there exists a set which consists solely of a single element taken from each set of S. It is of these axioms that Kurt Gödel said:

> Despite their remoteness from sense experience, we do have something like a perception also of the objects of set theory, as is seen from the fact that the axioms force themselves upon us as being true. I don't see any reason why we should have less confidence in this kind of perception, i.e., in mathematical intuition, than in sense perception ... They, too, may represent an aspect of objective reality.

other axioms of set theory (just as Euclid's parallel postulate was eventually shown to be independent of the other axioms of plane geometry) and therefore could be neither proved nor refuted from those axioms.

The lesson we learn here is that mathematical axioms are more like initial conditions for natural laws than we might have suspected. Indeed, it is the hope of some that they may turn out to be the same: that the ultimate assumptions one has to make about the input assumptions for the Theory of Everything are those required for logical consistency. But we have also learned that their nature and inter-relationship is extremely subtle. We are at liberty to choose whichever collection of them suits our purpose. Lacking an obvious intuitive guide as to the appropriateness of highly abstruse axioms (like the continuum hypothesis, for example), how do we know whether they should be included or not? Motivated by this experience with the continuum hypothesis problem Alonzo Church remarked that

> ...if a choice must in some sense be made among the rival set theories, rather than merely and neutrally to develop the mathematical consequences of alternate theories, it seems that the only basis for it can be the same informal criterion of simplicity that governs the choice among rival physical theories when both or all of them equally explain the experimental facts.

Cohen's demonstration that the continuum hypothesis is independent of the other axioms of set theory means that we are equally at liberty either to add it or its negation to the existing axioms of set theory. In each case, we could create a different enlargement of set theory, just as we can retain Euclid's parallel postulate or replace it by its negation to create logically consistent non-Euclidean geometries. If one is a mathematical Platonist who believes that mathematical entities really exist then only one of those two mutually exclusive set theories really exists, but if one is a constructivist or formalist then each are equally valid intellectual creations.

There is one area of interaction between fundamental physics and these foundational questions regarding infinity. It concerns the issue of whether a true continuum exists in reality or not. Most fundamental pictures of the physical world assume that the basic notions—fields, space, and time—are continuous entities rather than discrete bits. This issue of discreteness versus continuity is an ancient tension in natural philosophy that re-emerges in every era in new dress. The most important point about it for the structure of any theory of infinity is the vast difference in complexity that would exist between a continuous and a discrete Theory of Everything. The reason for this is that the number of *continuous* transformations that exist between one set of real numbers and another is a whole order of infinity lower than the total number of possible transformations which are not continuous. The requirement of continuity produces a vast and surprising reduction in scope. Since these continuous transformations include the catalogue of possible relationships from which we draw that class of transformations (or 'equations') called the laws of physics, we see that a discontinuous world will be infinitely more complex in its potentiality. It is less constrained in what it is allowed to do. At present, physicists are enamoured of symmetry and search only for continuous pictures of fundamental physics. Maybe, one day, they will be motivated to look at possible structures of a fundamentally discrete world. In Chapter 9, we shall look at some of the ideas that might provoke them to do so.

What statements can be proved or disproved depends crucially upon the information content of the axioms at hand. Some philosophers of science have used Gödel's theorems regarding the incompleteness of arithmetic (and hence of any logical system containing arithmetic) to argue that we can never know everything about the physical universe in terms of mathematical laws of Nature.

MATHEMATICAL JUJITSU

No mathematical theorem has aroused as much interest among non-mathematicians as Gödel's incompleteness theorem . . . One finds invocations not only in discussion groups dedicated to logic, mathematics, computing, or philosophy, where one might expect them, but also in groups dedicated to politics, religion, atheism, poetry, evolution, hip-hop, dating, and what have you.
— TORKEL FRANZÉN

Gödel's monumental demonstration, that complicated systems of mathematics have self-imposed limits on what they can prove, gradually changed the way in which philosophers and scientists viewed the world and our quest to understand it. Superficially, it appears that all human investigations of the Universe must be limited. Science is based on mathematics; mathematics cannot discover all truths; therefore science cannot discover all truths. This is an argument that is often heard. One of Gödel's contemporaries, the famous mathematician Hermann Weyl, described Gödel's discovery as exercising a 'constant drain on the enthusiasm' with which he pursued his scientific research. In more recent times, a frequent writer on theology and science,

Stanley Jaki, has argued that Gödel's theorem prevents us from gaining an understanding of the cosmos as a necessary truth,

> Clearly then no scientific cosmology, which of necessity must be highly mathematical, can have its proof of consistency within itself as far as mathematics goes. In the absence of such consistency, all mathematical models, all theories of elementary particles, including the theory of quarks and gluons ... fall inherently short of being that theory which shows in virtue of its a priori truth that the world can only be what it is and nothing else. This is true even if the theory happened to account with perfect accuracy for all phenomena of the physical world known at a particular time.

and so is a fundamental barrier to human understanding of the Universe:

> It seems on the strength of Gödel's theorem that the ultimate foundations of the bold symbolic constructions of mathematical physics will remain embedded forever in that deeper level of thinking characterized both by the wisdom and by the haziness of analogies and intuitions. For the speculative physicist this implies that there are limits to the precision of certainty, that even in the pure thinking of theoretical physics there is a boundary ... An integral part of this boundary is the scientist himself, as a thinker.

Intriguingly, and just to show the important role human psychology plays in assessing the significance of limits, some other scientists, like Freeman Dyson, acknowledge that Gödel places limits on our ability to discover the truths of mathematics and science, but interpret this as ensuring that science will go on forever. Dyson, who had some contact with Gödel during his time at Princeton's Institute for Advanced Study, sees the incompleteness theorem as an insurance policy against the scientific enterprise, which he admires so much, coming to a self-satisfied end; for

> Gödel proved that the world of pure mathematics is inexhaustible; no finite set of axioms and rules of inference can ever encompass the whole of mathematics; given any set of axioms, we can find meaningful mathematical questions which the axioms leave unanswered. I hope that an analogous situation exists in the physical world. If my view of the future is correct, it means that the world of physics and astronomy is also inexhaustible; no matter how far we go into the future, there will always be new things happening, new information coming in, new worlds to explore, a constantly expanding domain of life, consciousness, and memory.

In these two quite different statements, we see the optimistic and the pessimistic responses to Gödel. The optimists, like Dyson, see his result as a guarantor of the never-ending character of human investigation. They see scientific research as an essential part of the human spirit whose final completion would have a disastrous demotivating effect upon us. The pessimists, like Jaki, interpret Gödel as establishing that the human mind cannot know all of the secrets of Nature. They place more emphasis upon the possession and application of knowledge than on the process of acquiring it. The pessimist does not see the principal human benefit of science as arising from the quest for knowledge itself.

On reflection we should not be too surprised at such diametrically opposed responses. Many things in life create the same hiatus. It all depends whether you think your glass is half empty or half full. Gödel's own view was as unexpected as ever. He thought that intuition, by which we can 'see' truths of mathematics and science, was a tool that would one day be valued just as formally and reverently as logic itself,

> I don't see any reason why we should have less confidence in this kind of perception, i.e., in mathematical intuition, than in sense perception, which induces us to build up physical theories and to expect that future sense perceptions will agree with them and, moreover, to believe that a question not decidable now has meaning and may be decided in the future.

Gödel himself was not minded to draw any strong conclusions for the scope of physics from his incompleteness theorems. He made no connections with the Uncertainty Principle of quantum mechanics, which was another great deduction which limited our ability to know, and which was discovered by Heisenberg just a few years before Gödel proved his first theorem. In fact, Gödel was rather hostile to any consideration of quantum mechanics at all. Those who worked at the same Institute (no one really worked *with* him) believed that this was a result of his frequent discussions with Einstein who, in the words of John Wheeler (who knew them both) 'brainwashed Gödel' into disbelieving quantum mechanics and the Uncertainty Principle. Greg Chaitin records this account of Wheeler's attempt to draw Gödel out on the question of whether there is a connection between Gödel incompleteness and Heisenberg Uncertainty,

> Well, one day I was at the Institute for Advanced Study, and I went to Gödel's office, and there was Gödel. It was winter and Gödel had an electric heater and had his legs wrapped in a blanket. I said 'Professor Gödel, what connection do you see between your incompleteness theorem and Heisenberg's uncertainty principle?' And Gödel got angry and threw me out of his office!

The argument that mathematics contains unprovable statements, physics is based on mathematics, therefore physics will not be able to discover everything that is true, has been around for a long time. With these worries in mind, let us look a little more closely at what Gödel's result might have to say about the course of physics. The situation is not so clear-cut as the commentators would have us believe. It is useful to lay out the precise assumptions that underlie Gödel's deduction of incompleteness. Gödel's theorem says that if a formal system is

1. *finitely specified*
2. *large enough to include arithmetic*
3. *consistent*

then it is *incomplete*.

Condition 1 means that there is not an uncomputable infinity of axioms. We could not, for instance, choose our system to consist of all the true statements about arithmetic because this collection cannot be finitely listed in the required sense.

Condition 2 means that the formal system includes all the symbols and axioms used in arithmetic. The symbols are 0, 'zero', S, 'successor of', +, ×, and =. Hence, the number two is the successor of the successor of zero, written as the *term* SS0, and two and plus two equals four is expressed as $SS0 + SS0 = SSSS0$.

The structure of arithmetic plays a central role in the proof of Gödel's theorem. Special properties of numbers, like their primeness and the fact that any number can be expressed in only one way as the product of the prime numbers that divide it, were used by Gödel to establish the vital correspondence between statements of mathematics and statements about mathematics. Thereby, linguistic paradoxes like that of the 'liar' could be embedded, like Trojan horses, within the structure of mathematics itself. Only logical systems which are rich enough to include arithmetic allow these incestuous encodings of statements about themselves to be made within their own language.

Again, it is instructive to see how these requirements might fail to be met. Pick a theory that consists of references to (and relations between) only the first ten numbers (0, 1, 2, 3, 4, 5, 6, 7, 8, 9) with base-10 arithmetic, then Condition 2 fails and such a mini-arithmetic is complete. Arithmetic makes statements about individual numbers, or terms (like SS0, above). If a system does not have individual terms like this but, like Euclidean geometry, only makes statements about a continuum of points, circles, and lines, in general, then it cannot satisfy Condition 2. And so, as Alfred Tarski first showed, Euclidean geometry is complete. There is nothing magical about the flat, Euclidean nature of the geometry either: the non-Euclidean geometries on curved surfaces are also complete. Similarly, if we had a logical theory dealing with numbers that only used the concept of 'greater than' and 'less than' without referring to any specific numbers then it would be complete: we can determine the truth or falsity of any statement about real numbers involving the 'greater than' relationship.

Another example of a system that is smaller than arithmetic is arithmetic without the multiplication, ×, operation. This is called Presburger arithmetic (the full arithmetic is called Peano arithmetic after the mathematician who first expressed it axiomatically, in 1889). At first, this sounds strange—in our everyday encounters with multiplication it is nothing more than a shorthand way of doing addition (e.g., $2 + 2 + 2 + 2 + 2 + 2 = 2 \times 6$)—but in the full logical system of arithmetic, in the presence of logical quantifiers like 'there exists' or 'for any', multiplication permits constructions which are not merely equivalent to a succession of additions.

Gödel showed, as part of his doctoral thesis work, that Presburger arithmetic is complete: all statements about the addition of natural numbers can be proved or disproved; all truths can be reached from the axioms. Similarly, if we create another truncated version of arithmetic, which does not have addition, but retains multiplication, this is also complete. It is only when addition and multiplication are simultaneously present that incompleteness emerges. Arithmetic is the watershed in complexity for incompleteness to appear.

The use of Gödel's theorem to place limits on what a mathematical theory of physics (or anything else) can ultimately tell us seems at first to be a fairly straightforward consequence. But as one looks more carefully into the question, things are not quite so simple. Suppose, for the moment, that all the conditions required for Gödel's theorem to hold are in place. What would incompleteness look like in practice? We are familiar with the situation of having a physical theory which makes accurate predictions about a wide range of observed phenomena: we might call it 'the standard model' or 'string theory'. One day, we may be surprised by an observation about which it has nothing to say. It cannot be accommodated within its framework. Examples are provided by some so called 'grand unified theories' in particle physics. Some early editions of these theories had the property that all neutrinos must have zero mass. When neutrinos were observed to have a non-zero mass then we know that the new situation cannot be accommodated within our original theory. What do we do? We have encountered a certain sort of incompleteness, but we respond to it by extending or modifying the theory to include the new possibilities. Thus, in practice incompleteness looks very much like inadequacy in a theory.

In the case of arithmetic, if some statement about arithmetic is known to be undecidable (there are known statements of this sort, it means that both their truth and falsity are consistent with the axioms of arithmetic) then we have two ways of extending the structure. We can create two new arithmetics: one which adds the undecidable statement as an extra axiom, the other which adds its negation as a new axiom. Of course, the new arithmetics will still be incomplete, but they can always be extended to accommodate any incompleteness. Thus, in practice, a physical theory can always be enlarged by adding new principles which force all the undecidability into the part of the mathematical realm which has no physical manifestation. Incompleteness would then always be very hard, if not impossible, to distinguish from incorrectness or inadequacy.

An interesting example of this dilemma is provided by the history of mathematics. During the sixteenth century, mathematicians started to explore what happened when they added together infinite lists of numbers. If the quantities in the list get larger then the sum will 'diverge', that is, as the number of terms approaches infinity so does the sum. An example is the sum

$$1 + 2 + 3 + 4 + 5 + \ldots = \text{infinity.}$$

However, if the individual terms get smaller and smaller sufficiently rapidly* then the sum of an infinite number of terms can get closer and closer to a finite limiting value which we shall call the sum of the series; for example

$$1 + \frac{1}{9} + \frac{1}{25} + \frac{1}{36} + \frac{1}{49} + \ldots = \frac{\pi^2}{8} = 1.2337005.$$

This left mathematicians to worry about a most peculiar type of unending sum,

$$1 - 1 + 1 - 1 + 1 - 1 + 1 \ldots = \text{?????}$$

If you divide up the series into pairs of terms it looks like $(1-1) + (1-1) + \ldots$ and so on. This is just $0 + 0 + 0 + \ldots = 0$ and the sum is zero. But think of the series as $1 - \{1 - 1 + 1 - 1 + 1 - \ldots\}$ and it looks like $1 - \{0\} = 1$. We seem to have proved that $0 = 1$.

Mathematicians had a variety of choices when faced with ambiguous sums like this. They could reject infinities in mathematics and deal only with finite sums of numbers, or, as Cauchy showed in the early nineteenth century, the sum of a series like the last one must be defined by specifying more closely

---

* That the terms in the sum get progressively smaller is a necessary but not a sufficient condition for an infinite sum to be finite. For example, the sum $1 + \frac{1}{2} + \frac{1}{3} + \frac{1}{4} + \frac{1}{5} + \ldots$ is infinite.

what is meant by its sum. The limiting value of the sum must be specified together with the procedure used to calculate it. The contradiction $0 = 1$ arises only when one omits to specify the procedure used to work out the sum. In both cases it is different and so the two answers are not the same. Thus, here we see a simple example of how a limit is side-stepped by enlarging the concept which seems to create limitations. Divergent series can be dealt with consistently so long as the concept of a sum for a series is suitably extended.

Another possibility, which appears very likely to be true, is that the laws of Nature only use the decidable part of mathematics. We know that mathematics is an infinite sea of possible structures. Only some of those structures and patterns appear to find existence and application in the physical world. Very few of them are used to describe the laws of Nature. It may be that they are all from the subset of decidable truths.

It is also possible that the conditions required to prove Gödel's incompleteness do not apply to physical theories. Condition 1 requires the axioms of the theory to be listable. It might be that the laws of physics are not listable in this special sense. This would be a radical departure from the situation that we think exists, where the number of fundamental laws is believed to be not just listable, but finite (and very small). Yet, it is always possible that we are just scratching the surface of a bottomless tower of laws, only the top of which has significant effects upon our experience. However, if there were an unlistable infinity of physical laws then we would face a more formidable problem than that of incompleteness.

An equally interesting issue is that of finiteness. It may be that the universe of physical possibilities is finite, although astronomically large. However, no matter how large the number of primitive quantities to which the laws refer, so long as they are finite the resulting system of interrelationships will be logically complete. We should stress that although we habitually assume that there is a continuum of points of space and time this is just an assumption that is very convenient for the use of simple mathematics. There is no deep reason to believe that space and time are continuous, rather than discrete, at their most fundamental microscopic level; in fact, there are some theories of quantum gravity that assume that they are not. Quantum theory has introduced discreteness and finiteness in a number of places where once we believed in a continuum of possibilities. Curiously, if we give up this continuity, so that there is not necessarily another point in between any two sufficiently close points you care to choose, space-time structure becomes vastly more complicated. Many more complicated things can happen. This question of finiteness might also be bound up with the question of whether the

Universe is finite in volume and whether the number of elementary particles (or whatever the most elementary entities might be) of Nature are finite or infinite in number. Thus there might only exist a finite number of terms to which the ultimate logical theory of the physical world applies. Hence, it would be complete.

An interesting possibility with regard to the application of Gödel to the laws of physics is that Condition 2 of the incompleteness theorem might not be met. How could this be? Although we seem to make wide use of arithmetic, and much larger mathematical structures, when we carry out scientific investigations of the laws of Nature, this does not mean that the inner logic of the physical Universe needs to employ such a large structure. It is undoubtedly convenient for us to use large mathematical structures together with concepts like infinity but this may be an anthropomorphism. The deep structure of the Universe may be rooted in a much simpler logic than that of full arithmetic, and hence be complete. All this would require would be for the underlying structure to contain either addition or multiplication but not both. Recall that all the sums that you have ever done have used multiplication simply as a shorthand for addition. They would be possible in Presburger arithmetic as well. Alternatively, a basic structure of reality that made use of simple relationships of a geometrical variety, or which derived from 'greater than' or 'less than' relationships, or subtle combinations of them all could also remain complete although the proofs needed to demonstrate them become very long. The fact that Einstein's theory of general relativity replaces many physical notions like force and weight by *geometrical* distortions in the fabric of spacetime may well hold some clue about what is possible here.

The laws of physics might be fully expressible in terms of a mathematical system that is complete, but in practice we would always be far more concerned with making sure that we had got the *correct* system than a complete system. Tarski showed that, unlike arithmetic of natural numbers, the first-order theory of real numbers under addition and multiplication is decidable. This is rather surprising and may give some hope that theories of physics based on the real or complex numbers will evade undecidability. Many mathematical systems used in physics, like lattice theory, projective geometry, and Abelian group theory are also decidable, although others, notably non-Abelian groups are not.

There is another important aspect of the situation to be keep in view. Even if a logical system is complete, it always contains unprovable 'truths'. These are the axioms which are chosen to define the system and they are assumed to be independent of each other and consistent. And after they are chosen, all the logical system can do is deduce conclusions from them. In simple logical systems, like Peano arithmetic, the axioms seem reasonably obvious because we are thinking backwards—formalizing something that we have been doing intuitively for thousands of years. When we look at a subject like physics, there are parallels and differences. The axioms, or laws, of physics are the prime target of physics research. They are by no means intuitively obvious, because they govern regimes that can lie far outside our experience. The outcomes of those laws are unpredictable in certain circumstances because they involve symmetry breakings. Trying to deduce the laws from the outcomes is not something that we can ever do uniquely and completely by means of a computer program.

Thus, we detect a completely different emphasis in the study of formal systems and in physical science. In mathematics and logic, we start by defining a system of axioms and laws of deduction. Then, we might try to show that the system is complete or incomplete, and deduce as many theorems as we can from the axioms. In science, we are not at liberty to pick any logical system of laws that we choose. We are trying to find the system of laws and axioms (assuming there is one—or more than one perhaps) that will give rise to the outcomes that we see. It is always possible to find a system of laws which will give rise to any set of observed outcomes. But it is the very set of unprovable statements that the logicians and the mathematicians ignore—the axioms and laws of deduction—that the scientist is most interested in discovering, rather than simply assuming. The only hope of proceeding as the logicians do, would be if for some reason there is only one possible set of axioms or laws of physics which could include all the forces that we know of.

So, in summary so far, we have argued that there is no reason to believe that Gödel's incompleteness theorem places any restriction on our ability to find the ultimate laws of Nature—the Theory of Everything. Physics uses only a part of mathematics and that part can lie in the decidable area of mathematics. In fact, the mathematics used in expressing the known laws of Nature uses only simple patterns and the process of finding them is not beset by undecidability.

However, whereas the laws of Nature are simple, their outcomes are not. They are complicated and asymmetrical and we have often been in a situation where we have a law of Nature in the form of a system of equations but we don't know how to solve them to determine the outcomes of the laws.

It is in this realm of the complicated outcomes of the known laws that we expect Gödel incompleteness to rear its head. We already know of a number of questions that we could ask about the Universe that cannot be answered because of incompleteness. They are not restrictions on determining the laws of Nature but they do stop us using them to answer some simple questions that we might have expected to have accessible answers.

Specific examples have been given of physical problems which are undecidable. As one might expect from what has just been said, they do not involve an inability to determine something fundamental about the nature of the laws of physics or the most elementary particles of matter. Rather, they involve an inability to perform some specific mathematical calculation, which inhibits our ability to determine the course of events in a well-defined physical problem. However, although the problem may be mathematically well defined, this does not mean that it is possible to create the precise conditions required for the undecidability to exist.

An interesting series of examples of this sort have been created by the Brazilian mathematicians Francisco Doria and Newton da Costa. Responding to a challenge problem posed by the Russian mathematician Vladimir Arnold, they investigated whether it was possible to have a general mathematical criterion which would decide whether or not any equilibrium was stable. A stable equilibrium is a situation like a ball sitting in the bottom of a basin—displace it slightly and it returns to the bottom; an unstable equilibrium is like a needle balanced vertically—displace it slightly and it moves away from the vertical. When the equilibrium is of a simple nature this problem is very elementary; first-year science students learn about it. But, when the equilibrium exists in the face of more complicated couplings between the different competing influences, the problem soon becomes more complicated than the situation studied by science students. So long as there are only a few competing influences the stability of the equilibrium can still be decided by inspecting the equations that govern the situation. Arnold's challenge was to discover an algorithm which tells us if this can always be done, no matter how many competing influences there are, and no matter how complex their interrelationships. By 'discover' he meant find a formula into which you can feed the equations which govern the equilibrium along with your definition of stability, and out of which will pop the answer 'stable' or 'unstable'.

Strikingly, da Costa and Doria discovered that there can exist no such algorithm. There exist equilibria characterized by special solutions of mathematical equations whose stability is undecidable. In order for this undecidability to have an impact on problems of real interest in mathematical physics, the equilibria have to involve the interplay of very large numbers of different forces. While such equilibria cannot be ruled out, they have not arisen yet in real physical problems. Da Costa and Doria went on to identify similar problems where the answer to a simple question, like 'will the orbit of a particle become chaotic?', is undecidable.

The tentative conclusion we should draw from this discussion is that, just because physics makes use of mathematics, it is by no means obvious that Gödel places any straightforward limit upon the overall scope of physics to understand the laws of Nature of the Universe, but it will limit the sorts of questions we can answer about the details of their outcomes in practice.

INITIAL CONDITIONS AND TIME SYMMETRY

The historian is a prophet looking backwards.
— AUGUST VON SCHLEGEL

Sometimes initial conditions can exert such an all-pervasive influence that they create the impression that a new type of law is acting. The most familiar case is that of the so-called 'second law of thermodynamics' which stipulates that the entropy, or level of disorder, of a confined physical system cannot decrease with the passage of time. Thus, we see coffee cups breaking accidentally into pieces, but we never see a cup re-form from the fragments. Our desks naturally degenerate from order to disorder but never vice versa. However, the laws of mechanics that govern the manner in which changes can occur allow the time-reverse of each of these common motions. Thus a world in which china fragments coalesce into Staffordshire china cups and untidy desks evolve steadily into tidy ones violates no law of Nature. The reason that things are invariably seen to proceed from bad to worse in closed systems is because the starting conditions necessary to manifest order-increase are fantastically unusual and the probability that they arise in practice is tiny. The fragments of china would all need to be moving at precisely the right speeds and in just the right directions so as to convene to form a cup. In practice there are vastly more ways for a desk to go from order to disorder than from disorder to order. Thus, it is the high probability of realizing the rather 'typical' conditions from which disorder is more likely to ensue that is responsible for the illusion of a disorder-creating law of Nature.

This example of the second law of thermodynamics alerts us to the importance of understanding initial conditions, particularly in unfamiliar situations. For without that understanding we may be misled into seeking from a Theory of Everything explanations for things that it has no business explaining.

Moreover, we see how the choice (or accident) of initial conditions creates a sense of time directionality in a physical environment. The 'arrow' of entropy increase is a reflection of the improbability of those initial conditions which are entropy-decreasing in a closed physical system.

Everywhere we look in the Universe, we discern that closed physical systems evolve in the same sense from ordered states towards a state of complete disorder called thermal equilibrium. This cannot be a consequence of known laws of change, since at their most fundamental level these laws are time-symmetric—they permit the time-reverse of any allowed sequence of events. The initial conditions play a decisive role in endowing the world with its sense of temporal direction. In our later discussion of quantum cosmology, we shall explore some of the dramatic consequences of initial conditions for the entire Universe. It will become clear that some prescription for initial conditions is crucial if we are to understand the observed universe. A Theory of Everything needs to be complemented by some such independent prescription which appeals to simplicity, naturalness, economy, or some other equally metaphysical notion to underpin its credibility. The only radically different alternative would seem to lie in a belief that the type of mathematical description of Nature that we have come to know and love—that of causal equations with starting conditions—is just an artefact of our own preferred categories of thought and merely an approximation to the true nature of things. At a deeper level, a sharp divide between those aspects of reality that we habitually call 'laws' and those which we have come to know as 'initial conditions' may simply not exist.

TIME WITHOUT TIME

There is nothing new under the sun.
— ECCLESIASTES

Leibniz and Laplace both recognized a puzzling consequence of perfect determinism. If all our laws of motion are in the form of equations which determine the future uniquely and completely from the present, then by a perfect knowledge of the starting state it would be possible for a superbeing to predict the entire future history of the Universe from this raw material. Although the statements to this effect by Laplace are often quoted and the concept of determinism in classical physics has assumed the title 'Laplacian determinism', there is an earlier and more explicit statement of the idea in Boscovich's remarkable book of 1758 which we introduced in the last chapter. On the subject of determinism and continuity of motion, he writes:

> Any point of matter, setting aside free motions that arise from the action of arbitrary will, must describe some continuous curved line, the determination of which can be reduced to the following general problem. Given a number of points of matter and given, for each of them, the point of space that it occupies at any given instant of time; also given the direction and velocity of the initial motion if they were projected, or the tangential motion if they were already in motion; and given the law of forces expressed by some continuous curve [like his force law shown in Figure 2.1 of the last chapter] ... it is required to find the path of each of the points ... Now, although a problem of such a kind surpasses all the powers of the human intellect, yet any geometer can easily see thus far, that the problem is determinate ... a mind which had the powers requisite to deal with such a problem in a proper manner and was brilliant enough to perceive the solutions of it (and such a mind might even be finite, provided the number of points were finite, and the notion of the curve representing the law of forces were given by a finite representation), such a mind, I say, could from a continuous arc described in an interval of time, no matter how small, by all points of matter, derive the law of forces itself ... Now, if the law of forces were known, and the position, velocity and direction of all the points at any given instant, it would be possible for a mind of this type to forsee all the necessary subsequent motions and states, and to predict all the phenomena that necessarily followed from them.

Later the practicalities of attaining such perfect knowledge would be addressed by scientists, and then, in the twentieth century, the quantum theory would question the principle of whether such knowledge could be acquired by any observer, and indeed of whether it even exists in any meaningful sense. But let us leave aside these important developments and examine one of the striking consequences of the rigidly deterministic world of Boscovich, Laplace, and Leibniz that underpins the majority of the day-to-day concerns of physical scientists whose work is not directly affected by the ambiguities of quantum mechanics.

In a completely deterministic world, all the information about its structure is implicit in the initial conditions. The existence of time is a mystery. There is no use for it. Nothing really needs to 'happen' it all lies latent in the laws and initial conditions. A first reaction to this statement is to point to the laws of Nature as being algorithms that predict the future from the past, but we have seen that laws are equivalent to invariance principles, that is, statements to the effect that some entity does not change. The deterministic straitjacket makes time appear superfluous. Everything that is ever going to happen is implicit in the starting state. Our present state contains all the information necessary to reconstruct the past and predict the future. In Joseph Conrad's disquieting words from the *Heart of Darkness*,

> The mind of man is capable of anything—because everything is in it, all the past as well as all the future.

This situation always presented scientists of the pre-quantum era with a dilemma. During the nineteenth-century debates about the likelihood of Darwinian evolution as opposed to a special creation of the living world in its present wondrously adapted form, several scientific commentators remarked upon the essential convergence of these two views since the present state of the evolved world can be nothing more nor less than a precise mirror of particular initial conditions. Others fretted over the problem of free will in a world of rigid determinism. A consideration of this problem led James Clerk Maxwell to appreciate the world of difference between determinism in principle and determinism in practice.

There exist a vast number of physical situations, from the weather to a beating heart, where the slightest uncertainty in our knowledge of the state of the system at one moment results in total loss of information about its exact state after a very short period of time. Almost identical presents lead to very different futures. Such systems are called 'chaotic'. Their prevalence is responsible for many of the complexities of life: the economy, money-market fluctuations, or climatic variations. In these situations, it does not matter how precisely we may know the rules governing how changes occur because we cannot ascertain the present state of things with perfect accuracy. Our capacity to predict rapidly becomes empty. It is curious how long it took scientists to recognize the overwhelming influence of such sensitivity to starting conditions in the real world. So blinkered were they by the deterministic clockwork of the Newtonian world-view and the technological advances that grew out of it that the 'laws that never shall be broken' stood out as the dominant aspect of the world's character. Only the deepest thinkers of the nineteenth century, like Maxwell and Poincaré, recognized the true nature of things, which so often leaves us unable to predict the actual future even if we had the precise laws of Nature in our hands. Maxwell's thoughts about the problem of free will in practice led him to recognize that many sequences of natural events possess an extremely sensitive dependence upon their precise starting conditions. Later, it was Henri Poincaré's attempts to understand the sensitive dynamics of the planetary motions in our solar system that led him also to appreciate that a very small cause which escapes our notice determines a considerable effect that we cannot fail to see, and then we say that the effect is due to chance. If we knew exactly the laws of nature and the situation of the universe at the initial moment, we could predict exactly the situation of the same universe at a succeeding moment. But even if it were the case that the natural laws had no longer any secret for us, we could still only know the initial situation *approximately*. If that enabled us to predict the succeeding situation with *the same approximation*, that is all we require, and we should say that the phenomenon had been predicted, that is, governed by laws. But it is not always so; it may happen that small differences in the initial conditions produce very great ones in the final phenomena. A small error in the former will produce an enormous error in the latter. Prediction becomes impossible, and we have the fortuitous phenomenon.

Here, Poincaré points out that this extreme sensitivity that the evolution possesses to the actual state of the motion leads to very complicated and erratic behaviour that cannot be uniquely traced back through its antecedents in practice. Hence, it is regarded as a 'random' phenomenon by those who observe it. There is nothing intrinsically indeterminate about the motions involved. If we could have perfectly accurate knowledge of the starting conditions, we could predict the future behaviour perfectly. What we now know that Poincaré did not is that quantum aspects of reality forbid the acquisition of such error-free knowledge of the initial conditions *in principle*, not merely in practice. Nor are these quantum restrictions far removed from experience. If we were to strike a snooker ball as accurately as the quantum uncertainty of Nature permits, then it would take merely a dozen collisions with the sides of the table and other balls for this uncertainty to have amplified to encompass the extent of the entire snooker table. Laws of motion would henceforth tell us nothing about the individual trajectory of the ball.

Before leaving these prescient remarks of Maxwell and Poincaré, it is intriguing to search in Boscovich's work to find his thoughts about the practicalities of some 'mind' grasping the content of all motions. He seems to recognize the inevitability of perturbing influences in reality, although not their unstable character. And, of any aspiration to exploit determinism to obtain complete knowledge, he cautions:

We cannot aspire to this, not only because our human intellect is not equal to the task, but also because we do not know the number, or the position and motion of each of these points . . . and there is another reason namely that the free motions produced by spiritual substances affect these curves . . .

The ubiquity of chaotic phenomena raises a further problem for our dreams of omniscience through the medium of a Theory of Everything. Even if we can overcome the problem of initial conditions to determine the most natural or uniquely consistent starting state, we may have to face the reality that there is inevitable uncertainty surrounding the prescription of the initial state which makes the prediction of the exact future state of the Universe impossible. Only statistical statements will be possible.

COSMOLOGICAL TIME

Time is God's way of keeping things from happening all at once.
— ANONYMOUS TEXAN GRAFFITI

In most scientific problems the initial conditions are rather mundane. We prepare them in a particular way in order better to watch a certain type of effect which we suspect will ensue. But in cosmology—the study of the structure and evolution of the Universe as a whole—the situation is altogether more interesting. For, without some knowledge of those cosmic initial conditions, our knowledge of the Universe remains seriously incomplete. It would appear that even knowledge of the Theory of Everything would prevent us understanding why the Universe began in a particular way. Given a sequence of numbers, we might guess the pattern between them which allows the next one to be predicted and the whole sequence to be algorithmically compressed, but be unable to say why it begins at the particular point that it does. Yet, what really singles out the problem of cosmological initial conditions is that it has metaphysical consequences. If there are special initial conditions which start the evolution of the Universe upon the course that leads to the present, what is it that selects those rather than any other starting conditions?

Initial conditions determine the coarse-grained structure of the Universe over its largest dimensions. They will play a role in determining the size of the Universe, its shape, its temperature, and its composition. From what we have already said the situation appears clear-cut. There will be particular initial conditions which lead to the present observed state. All we can hope for is to discover what they were. But we shall see that the situation is more interesting than that, and for over twenty-five years the attitude of cosmologists to the issue of initial conditions has fuelled almost all our ideas about the structure of the Universe. And, because those initial conditions were set up more than ten billion years ago when the Universe resembled a vast experiment in high-energy physics, their consideration brings cosmology into collision with our thinking about the ultimate structure of the elementary particles of matter. The question of why the Universe is as it is, is inextricably linked to that of why fundamental physics is the way that it is.

Let us begin by exploring the ramifications and options of the traditional cosmological pictures in which there is a fundamental distinction between the laws of Nature and initial conditions.

After Hubble's discovery in the late 1920s that the Universe is in a state of overall expansion it was appreciated that this implies that the Universe must have had a 'beginning' in the sense that the present state of expansion could not be extended indefinitely into the past. We appear to encounter a moment in our finite past when the density was infinite and all matter was squashed to zero size. Later, in the mid-1960s, this 'Big Bang' picture was reinforced by the discovery of a cosmic heat radiation field, greatly cooled by the expansion, which had been predicted should exist as a remnant of the early hot state. Subsequently, the careful study of the expanding universe models supplied by Einstein's theory of general relativity has confirmed further detailed predictions based upon what the Universe must have been like when it was just one second old. It is generally agreed by modern cosmologists that we have established the general framework of how the Universe behaved from when it was a second old until the present, some fifteen billion years later. This is not to claim that we understand everything that occurred. We do not understand the detailed processes by which galaxies formed, but such processes actually exert a negligible influence upon the course of the overall expansion. Prior to one second after the apparent beginning, we are on altogether shakier ground. We no longer have direct fossil remnants from the early universe against which to check the accuracy of our reconstruction of its history. In order to reconstruct the history of the Universe in these first instants, we require knowledge of the behaviour of matter at far higher energies than are accessible to us by terrestrial experiments. Indeed, the study of the very early stages of the Universe's history may be the only way in which we can test our theories about the behaviour of matter at very high temperatures. For we might find that if a certain hypothetical elementary particle were really to exist then it would survive the Big Bang in such profusion that the strength of its gravitational pull today would have caused the Universe's expansion to decelerate at a rate far in excess of what is observed.

We are therefore caught in a double bind. We need to know the behaviour of the elementary particles of matter in order to understand the very early universe, but we need to know what the early universe was like in order to discover the behaviour of elementary particles.

With this warning taken on board, let us none the less continue to extrapolate our successful picture of the Universe into the first second of its history using the latest ideas in elementary-particle physics as a guide to what is possible or probable during the dim and distant past.

Traditionally (and currently) there are three distinct attitudes towards the problem of cosmological initial conditions:

- Show there are none.
- Show that their influence is minimal.
- Show that they have a special form.

The first option springs from a belief that the universe did not have a beginning—that there was no initial state. This stance was taken most adamantly by Hermann Bondi, Fred Hoyle, and Thomas Gold, who introduced the 'steady-state' theory of the Universe in 1948. The specific theory they proposed fell into conflict with observation long ago and its specific details are not important for our present discussion. What is most interesting is their motivation to avoid any special times occurring during the history of the Universe, just as Copernicus cautioned us against endowing special significance upon any places in the Universe. Clearly, if the Universe begins expanding (or existing) at some finite past moment or ceases to expand (or exist) at some future moment, then these moments are special times for any observer. The 'steady-statesmen' called the extension of the Copernican Principle from spatial location to spatial *and* temporal location, the *Perfect Cosmological Principle* (a title which provoked Herbert Dingle into remarking that this was like 'calling a spade a perfect agricultural instrument' and some Americans to suggest that the stipulation that the Universe be the same at all times was merely a device by which its proposers could ensure that there would always be an England). Although the steady-state universe expands, it maintains a constant density at all times by the assumption that matter is being continuously created at a rate that exactly counterbalances the rarefaction that would otherwise result from the expansion. This continuous creation contrasts with the once and for all creation that was envisaged in the Big Bang cosmological models of that time. The fact that the creation rate exactly balances the effects of the expansion was automatically ensured and the creation rate is so tiny, less than one atom in a cubic metre every ten billion years, that it could not be detected directly.

Yet, despite the fact that there is no actual beginning to the Universe in this theory—it always has, and always will expand, on the average, at the same constant rate—it still requires its defining parameters to be specified: there is no unique steady-state universe. The value of its constant universal density of matter, or, equivalently, its constant creation rate or the universal rate of expansion, needs to be explained. We must specify certain conditions at some moment of time to define this model. A Theory of Everything might tell us that the Universe has no beginning in time and expands in a fashion similar to the steady-state universe (at least until about ten billion years ago), but this would leave many things unexplained: the expansion rate of the Universe, the origin of galaxies, the heat content of the Universe, its imbalance between matter and antimatter.

This logical incompleteness characterizes any cosmological model that is hypothesized to have existed from a past infinity of time. It still requires extra specifications that play the role of 'initial' conditions, even if there is strictly no 'initial' moment in the temporal sense. In an infinitely old universe, initial conditions are required at past temporal infinity.

It is interesting to reflect that for centuries philosophers and theologians have attempted to settle by pure thought the issue of whether the Universe could or could not be infinitely old. That is, some have attempted to show that there is some logical contradiction inherent in the notion of a past infinity of time. And some still do. Such ideas have some association with cosmological arguments for the existence of God, which not only seek to demonstrate that there must have been an origin to the Universe in time but go further in showing (or, in practice, assuming) that this requires there to have been an originator. This is a slippery argument, notwithstanding our ultimate ignorance about such overwhelming questions. A common form of this argument points to the fact that everything that we see has a cause, and hence the Universe must have a cause. But this argument has a dangerous bend in the middle of it. The Universe is not a 'thing' in the sense of all the other examples that are being cited. It is a collection of things, or as Wittgenstein put it 'the world is the totality of the facts'. Our argument is thus seen to be analogous to arguing that all members of clubs have mothers, and therefore all clubs have mothers. One might also take issue with the claim that all events have causes. In the shadowy world of quantum theory, this need no longer be the case. We cannot tie individual observations to specific causes according to some interpretations of the quantum theory, and indeed this is one of the reasons why a quantum description of the whole Universe can in principle give a description of the creation of the material Universe without any direct initial cause being invoked.

When discussing those features of the steady-state universe which would have to be specified to complement what the laws of Nature tell us, we mentioned the expansion rate, but not the shape, of the Universe. One of the unusual features of the steady-state model was that it was stable against any influences that might distort it away from possessing the same rate of expansion in every direction of the sky. If some violent event suddenly occurred somewhere in the Universe or the Deity temporarily intervened at one moment to make it expand faster in one direction than another, then with the passage of time these deviations from the state demanded by the Perfect Cosmological Principle would soon fade away and the expansion settle back into a perfectly symmetrical state. Such a property is a very attractive feature of any cosmological model because our Universe is observed to expand at the same rate in every direction to within one part in a thousand. The wider quest for a natural explanation of this surprising fact brings us to the second of the three general approaches that have been made to understanding the initial conditions of the Universe.

It is evident that the most awkward feature about the influence of initial conditions in cosmology is the fact that they are the most uncertain aspect of our knowledge. It may well be that we can never know how (or if) the Universe began. Therefore there has always been a lobby of cosmological opinion that has seen it as expedient to seek an explanation for the present structure of the Universe that places the minimum onus for that structure upon those unknowable initial conditions. But how could this be done?

There are many physical systems which rapidly lose memory of their initial conditions. By this we mean that their future states are to very high accuracy pretty much the same regardless of how they started out. Stir a large pot of treacle in a vigorous way and it will quickly settle down to the same placid state no matter how you stirred it. Drop a rock in air from a sufficiently great height and it will hit the ground at essentially the same speed no matter how hard you threw it initially because the competing effects of gravity accelerating the stone and air resistance slowing it down always act to create a situation where they have an equal and opposite effect, and thereafter the stone feels no net force at all and falls at constant speed. The Universe could be like this. Cosmologists spent much of the 1970s looking for natural physical processes which might emerge during the early stages of the Universe and render its present state quite inevitable irrespective of the details of how it started. In particular, they hoped to explain why the visible universe possesses the remarkable property of expanding at the same rate in every direction to within one part in ten thousand. If it could be shown that no matter how disparate were the expansion rates in different directions when the Universe began, so long as we wait long enough (and life takes a long time to evolve), we will always find almost identical expansion rates in different directions because physical processes always arise to transport energy from place to place and iron out disparities in expansion energy between one direction and another.

This sounds like an attractive scenario. Unfortunately, the early attempts to implement it were largely unsuccessful. The main problem is that the smoothing of irregularities is one of those processes that is governed by the second law of thermodynamics. Irregularity in the expansion can only be reduced if this partial reduction in disorder (or 'entropy' as it is called) is paid for by an even larger production of entropy in another form. In practice, this compensating entropy appears in the form of heat radiation. Thus, if we build a chair out of disordered pieces of wood, we do not violate the second law, because we put a lot of physical and mental effort into it, which is manifested as the production of heat and sound by our bodies. Yet we find that the Universe does not contain very much heat radiation today and therefore very little smoothing of irregularities can have occurred in the past. Moreover, even if smoothing were to occur, there exists a vast array of cosmological models in which the irregularities could not become smooth by the present day. The smoothing effects are not strong enough to overcome a tendency to become increasingly distorted that is latent in the starting conditions of some possible universes.

As a result of these negative discoveries, cosmologists had become somewhat disenchanted with this route to explaining the large-scale regularity of the Universe by the end of the 1970s. But then a new idea emerged. Alan Guth pointed out that if the expansion rate of the Universe could be greatly increased for a short period during its early stages then one could explain the present structure of the Universe with only a minimal appeal to initial conditions, without having to worry about producing excessive heat.

The inflationary universe is a recipe for doing just this. It is based upon the expectation that there exist certain types of matter in the realm of elementary particles which, in effect, behave as though they exhibit gravitational repulsion rather than attraction. This is possible because they possess a negative pressure, or tension, and in the theory of relativity all forms of energy—and pressure is one of them—feel the force of gravity since they are all equivalent to a mass (via Einstein's famous $E = mc^2$ formula relating energy E to mass m and the speed of light c). If such a tension can appear during the earliest moments of the Universe's expansion, then gravity no longer pulls matter back and decelerates the expansion of the Universe. Instead, it acts to *accelerate* the expansion. The period of acceleration is called the *inflation* of the Universe. It causes all distorting influences to diminish extremely quickly and the Universe rapidly assumes a highly symmetrical state of expansion, which explains the residual state of extreme regularity that we still witness today. If the period of inflation lasts for only a very brief period, it is sufficient to reduce all irregularities that might have been present initially to an infinitesimally small level. It wipes the slate clean. Thus, it claims to explain the regular expansion that we presently observe, irrespective of the initial conditions. This is actually not quite true. There are always some maliciously chosen initial conditions that will not be damped down sufficiently by a *pre-specified* period of inflation, but, conversely, if the initial conditions are chosen first, then there always exists an amount of inflation that will suffice. It is something of a chicken-and-egg problem. If you are allowed to pick the period of inflation after you have picked the initial conditions then you can always explain what we see, but if the period of inflation is fixed first by the laws and constants of Nature then there are always initial conditions whose influence cannot be made innocuous by the present. The answer to the question 'What should be chosen first?' depends in a deep way upon one's view of initial conditions and their relationship with the laws of Nature. If we retain the traditional classical view that initial conditions are independent of the laws of physics, then, in the absence of other information to the contrary, we should regard the initial conditions of the Universe as being freely specifiable, but they would then possess a secondary status with respect to the laws and the constants of physics. We can envisage different initial conditions quite easily and are accustomed to specify them at will every time we employ laws of physics in the laboratory, but to alter a law of physics or the value of a fundamental constant is altogether more radical. Thus, it seems most reasonable to regard the constants and laws of physics, and hence the duration of any period of inflation, as having been fixed *before* we specify initial conditions. With this choice, inflation cannot always deliver the observed Universe irrespective of initial conditions. It might still turn out that the unsuccessful starting states are in some sense 'unlikely' ones, but the question of what distinguishes a probable from an improbable initial state is still an open one.

In the traditional Big Bang picture of the expanding universe, the relative sameness of the observed universe from place to place is something of a mystery. To understand the mystery, we must first distinguish between the entire Universe, which might be infinite in extent, and the 'visible universe', which is that part of it from which light has had time to travel since the expansion began. The visible universe can be thought of as a sphere of radius approximately equal to fifteen billion light years centred upon us. Fifteen billion light years is the distance that light can have travelled in the fifteen billion years we shall use as a good estimate of the time that has passed since the expansion apparently began. (It is a reasonable average of the different pieces of observational evidence which point to a 13 to 18 billion year age range for the universal expansion.) We know nothing about the Universe except from what we observe of the finite visible portion of it. For instance, no observations of the visible universe can ever tell us whether the entire Universe is finite or infinite.

It is our visible universe which displays remarkable large-scale uniformity. Yet, if we extrapolate this visible region backwards in time, we can determine how much smaller it would have been at earlier times in the Universe's history. For example, when the Universe was one second old, our present visible universe would have been crammed into a region only one and a half light years in size. When the Universe was 10^{-35} of a second old, it would have been squeezed into a region a mere centimetre across. This sounds staggeringly small, but for the cosmologist it is unacceptably large. It is 3×10^{25} times larger than the size of the regions whose contents are in causal contact at that early time: for at that time the latter distance is simply 10^{-35} seconds multiplied by the speed of light (3×10^{10} centimetres per second), which gives 3×10^{-25} centimetres. The upshot of this state of affairs is that the region which grows into the entire visible universe today is composed of a vast number of totally independent regions that cannot even 'know' of each other's existence at very early times (see Figure 3.2).

The root of this 'horizon problem', as it is called, is apparent from our description. The Universe expands too slowly early on, so that part of it

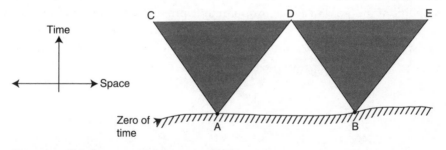

Figure 3.2 Signals sent out from two separate points, A and B, when the Universe begins expanding cannot reach each other until the time D. The interior of the wedges CAD and DBE represent the parts of space and time that can be contacted by signals emanating from A and B, respectively. This restriction upon communication arises because signals cannot travel faster than the speed of light; this means that communication is confined to the interior of the wedges. Notice that A cannot predict the future. Conditions at D are not determined solely by the signal that A transmits, but also by that transmitted from B.

which will encompass our visible universe today needs to have grown from a relatively large region at early times, a region far larger than any that can be kept smooth and regular at those times by physical processes that are limited in the extent of their influence by the speed of light. However, the period of accelerated expansion that characterizes the early evolution of the inflationary-universe models enables our entire visible universe to have evolved from a much smaller region at an early time like 10^{-35} of a second. In fact, if the inflation lasted for just a fleeting moment—from 10^{-35} to 10^{-33} of a second—then our entire visible universe can have emerged from a region that was within the range of light signals at these very early times. The gross uniformity of the observed universe now has a plausible explanation. It is the expanded image of a minute region that was small enough to have been smoothed by physical processes obeying the restrictions on their scope imposed by relativity.

In the standard Big Bang theory in which inflation does not occur, the observed universe cannot have arisen from any such causally correlated and coherent region. Instead, it is the coming together of a myriad of completely unrelated regions that would be expected to be very different from one another and hence result in a visible universe that was wildly different from place to place.

This new picture of the early evolution of the Universe radically diminishes the role of initial conditions, because, although the entire visible universe partially reflects the structure of some 'initial' conditions that define the structure of the Universe prior to the onset of the inflation, the particular initial conditions that play that role are only a minute part of the entire map of initial conditions for the whole (possibly infinite) Universe (see Figure 3.3).

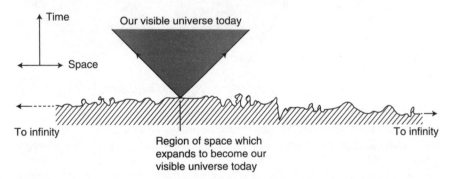

Figure 3.3 The structure of the visible universe is determined by conditions over only a tiny part of the 'initial' conditions of the Universe. If the Universe is infinite in size, then both our visible part of it and the part of the initial conditions which determine it are but infinitesimal parts of the whole.

This is disturbing to the scientist. It means that our observations of the structure of the visible universe can at best give us information about only a minute part of the initial conditions characterizing the first moments of the expanding universe. We can never know about the structure of the whole of the initial conditions for the Universe by observational science. They are condemned to remain always partially within the realm of philosophy and theology. It also makes any test of this theory extraordinarily difficult. Even if some varieties of initial condition do not allow any or enough inflation to take place, there will always be some part of the entire Universe initially where acceptable conditions will exist and that is all we require. As we shall see in a later chapter, we need to explore in some detail how our own existence plays a role in evaluating such theories.

The picture of initial conditions that inflation presents us with is therefore of a possibly chaotic or random initial state for the Universe as we look from place to place—rather like the surface of the sea. Each minute local region will inflate independently of all the others by an amount determined by its local conditions. We will find ourselves living inside one of these regions after it has greatly expanded. The inside of this region should look very smooth and expand uniformly, but beyond its boundary there are regions whose light rays have not yet had time to reach us. And these regions beyond our ken will in all probability be utterly different in structure. We have a picture which can explain why our visible part of the Universe is smooth even though the entire Universe would not be expected to be.

The inflationary period of expansion does not smooth out irregularity by entropy-producing processes like those explored by the cosmologists of the seventies. Rather, it sweeps the irregularity out beyond the horizon of our visible universe, where we cannot see it. The entire universe of stars and galaxies on view to us, on this hypothesis, is but the reflection of a minute, perhaps infinitesimal, portion of the Universe's initial conditions, whose ultimate extent and structure must remain forever unknowable to us. A Theory of Everything does not help here. The information contained in the observable part of the Universe derives from the evolution of a tiny part of the initial conditions for the entire Universe. The sum total of all the observations we could possibly make can only tell us about a minuscule portion of the whole.

It is possible that the rigid divide between laws and initial conditions that we have just assumed does not exist; that for some laws only one type of initial condition is allowed. This is a possibility that we shall now explore a little further.

The last of our options regarding the initial conditions of the Universe that we have left to consider is that there exists some special type of initial condition—effectively, a 'meta-law' governing initial conditions. The inflationary philosophy chooses to regard the initial conditions as being freely specifiable and inflation is a means by which we can show that their precise form has little bearing on what we will see today, so long as the laws of physics, the nature of elementary particles, and the constants of Nature permit this magical phenomenon of inflation to occur. By contrast, the lobby for special initial conditions searches for a fundamental link between the notion of laws and initial conditions that transcends our normal experience in classical physics. Traditionally, initial conditions are not constrained by the form of the laws of change except in a very weak fashion. If a solution of an equation of change also fixes the starting conditions uniquely, then this invariably means that the solution in question is extremely special, and hence unlikely to be realized in practice. To find a deep connection between the form of laws of Nature and their permitted starting conditions, we need therefore to look to a situation where there exists some probabilistic element regarding the possible form of evolutionary behaviour. This is something that can be found in any quantum description of things. For the most part, these attempts to link laws with initial conditions have focused upon the rapidly growing, embryonic subject of quantum cosmology. In so doing, they find themselves embroiled in other deep problems of a fundamental nature regarding the interpretation of quantum theory, about which much has been written elsewhere, and the less frequently discussed *problem of time*.

THE PROBLEM OF TIME

The English are not a very spiritual people. So they invented cricket to give them some idea of eternity.
— GEORGE BERNARD SHAW

There is a long-standing philosophical puzzle regarding the nature of time that has emerged in the works of different thinkers over millennia. It reduces to the question of whether time is an absolute background stage on which events are played out but yet remains unaffected by them, or whether it is a secondary concept wholly derivable from physical processes and hence affected by them. If the former picture were adopted, then we could talk about the creation of the physical Universe of matter *in time*. It would be meaningful to discuss what occurred before the creation of the material universe and what might happen after it passed away. Here, time is a transcendent part of reality without a conceivable beginning or end. This idea lends itself readily to the Platonic notion that there exist certain eternal truths or blueprints from which the temporal realities derive their qualities. Indeed, time takes upon itself many of the qualities traditionally associated with a Deity. The alternative, an idea that emerges in Aristotle's writings and more memorably in those of Augustine and Philo of Alexandria, before being elaborated by some of the early Islamic natural philosophers, is that time is something that comes into being with the Universe. Before the Universe was, there was no time, no concept of 'before'. Such a device enabled the medieval Scholastics to evade difficult conundrums about what took place before the creation of the world and what the Deity was doing in that period. In essence, this views time as a derived phenomenon, inextricably bound up with the contents of the Universe. The beginning of time is the moment when constants and laws of Nature must come into being ready-made and ready to go. In *The City of God*, St Augustine writes:

> Then assuredly the world was made, not in time, but simultaneously with time. For that which is made in time is made both after and before some time—after that which is past, before that which is future. But none could then be past, for there was no creature, by whose movements its duration could be measured. But simultaneously with time the world was made.

This is close to our common experience of time. We measure time using clocks, which are made of matter and which obey laws of Nature. We exploit the existence of periodic motions, whether they be revolutions of the Earth, oscillations of a pendulum, or vibrations of a caesium crystal; and the 'ticks' of these clocks define the passage of time for us. We have no everyday meaning to give to the notion of time aside from the process by which it is measured. We might thus defend an operationalist view, wherein time is defined by its mode of measurement alone.

Whereas, on the transcendental view of time, we might speak of bodies moving *in* time, the emphasis of the latter view is upon time being defined by the motion of things. One of the advantages of the first view is that one knows where one stands and what time is always going to look like: it is the same yesterday, today, and forever. By contrast the second picture promises to produce novel concepts of time—and might even do away with the concept altogether—as the material contents of the Universe alter their nature under varying conditions. We should be especially conscious of such a possibility as we backtrack towards those moments of extremis in the vicinity of the Big Bang. For any moment that appears to be the beginning of time inevitably exists where the very notion of time itself is likely to be most fragile. In an expanding and constantly changing universe, the operational view of time is likely to produce a subtle and variable conception of time's place and meaning.

ABSOLUTE SPACE AND TIME

I do not define time, space, place and motion, as being well known to all. Only I must observe, that the common people conceive those quantities under no other notions but from the relation they bear to sensible objects. And thence arise certain prejudices...
— ISAAC NEWTON

The image of a transcendent absolute time shadowing the march of events upon a cosmic billiard table of unending and unchanging space was the foundation of Newton's monumental description of the world. Once the equations governing the change of the world in space and time are given then the whole future course of events is determined by the starting conditions.* Time appears superfluous. Everything that is going to happen is programmed into the starting state.

The Newtonian laws of motion could be applied to the description of the world and followed backwards in time. Our Universe is observed to be expanding, and hence a Newtonian description leads to the assertion that there must have been a past moment of time at which everything was compressed to zero size and infinite density, the 'Big Bang' as it was first termed by Fred Hoyle. However, because of the absolute nature of space and time in the Newtonian world-view, we cannot draw any conclusions about the Newtonian Big Bang constituting an origin to time, let alone the origin of the Universe. It is simply a past time at which known laws predict that some physical quantities become unboundedly large; we say they become infinite in value there. But space and time go on regardless.

* This will not be true if other physical processes become involved. For example, in the archetypal situation of billiard balls moving according to Newton's laws, their future behaviour after collisions depends upon the rigidity of the collisions and this involves knowledge of the behaviour of the materials out of which the balls are made. This information is beyond the scope of Newtonian mechanics.

The first scientists to contemplate the significance of places where things apparently cease to exist or become infinite ('singularities', as we would now call them) in Newtonian theory were the eighteenth-century scientists Leonhard Euler and Roger Boscovich. They both considered the physical consequences of adopting force laws for gravitation other than Newton's famous inverse-square law. They found some of the alternatives had the unpleasant feature that the solutions just cease to exist after some definite time in the future when one studied the behaviour of objects orbiting around a central sun. They cannot be continued forwards any further in a world governed by one of these maverick force laws. Boscovich thinks it absurd that the body must disappear from the Universe at the centre if the force law were inverse cube rather than inverse square. He draws attention to Euler's earlier study of motion under the influence of gravity, where the master-mathematician

> asserts that the moving body on approaching the centre of forces is annihilated. How much more reasonable would it be to infer that this law of forces is an impossible one?

These appear to be the first contemplations of such matters in the context of Newtonian mechanics.

In fact, there are deep problems with attempting to apply Newton's theory of gravity and motion to the Universe as a whole. It will not tolerate the consideration of an *infinite* space distributed with matter: this leads to an infinite aggregate of gravitational influences at any one point due to the infinite number of gravitational attractions exerted by the others. Therefore a Newtonian universe must be finite in size and hence possess a boundary in space. If we think of Newtonian space stretching out straight in every direction, then this boundary must be a definite edge. For example, if the space is spherical about us at the centre, then the surface of the space is the surface of the sphere. Alternatively, the spatial universe could be a cube whose boundary was composed of the six faces of the cube. This prospect of a Universe with boundaries is a rather unattractive picture because we must specify how all physical quantities behave at these boundaries when the Universe is started at some time in the past. Thus, the Newtonian world requires the universe of matter to be a finite island of matter in an ocean of infinite absolute space.

Worse still, Newton's theory is incomplete. It does not contain enough equations to tell us how all the allowed changes to the Universe actually occur. If the Universe expands or contracts at exactly the same rate in every direction then everything is indeed determined, but when any deviations from perfectly spherical expansion are allowed at the start then determinism breaks down for there are no Newtonian laws which dictate how the shape of the world will change with time. Clearly, Newton's theory of absolute space and time is defective. The next step to take is to contemplate some coupling of the notions of space and time to the material contents of the world.

The earliest and most intriguing speculation of this sort was made by William Clifford, an English mathematician who contemplated just the type of situation that Einstein would build into the general theory of relativity. Clifford was motivated by the mathematical investigations of Riemann who had formalized the geometric study of curved surfaces and spaces which possess non-Euclidean geometry (that is, the three interior angles of a triangle no longer add up to 180 degrees where the three corners of the triangle are formed by joining the shortest lines that can be drawn between them to form the sides of the triangle on the curved surface). Clifford appreciated that the traditional space of Euclid is thus one of many and we can no longer assume that the geometry of the real world possesses the simple Euclidean form. The fact that it appears to be flat locally is not persuasive because most curved surfaces appear flat when viewed over small areas. After studying Riemann's ideas, Clifford proposed the following radical scenario in his paper of 1876:

> I wish here to indicate a manner in which these speculations may be applied to the investigation of physical phenomena. I hold in fact
>
> (1) That small portions of space *are* in fact of a nature analogous to little hills on a surface which is on the average flat; namely, that the ordinary laws of geometry are not valid in them.
>
> (2) That this property of being curved or distorted is continually being passed on from one portion of space to another after the manner of a wave.
>
> (3) That this variation of the curvature of space is what really happens in that phenomenon which we call the *motion of matter*, whether ponderable or etherial.
>
> (4) That in the physical world nothing else takes place but this variation, subject (possibly) to the law of continuity.

This prescience is rather remarkable. Although Einstein never seems to have been aware of these remarks, Clifford's intuitive idea became the central idea of the general theory of relativity. The geometry of space and the rate of flow of time are no longer absolutely fixed and independent of the material content of space and time. The matter content and its motion determine the geometry and the rate of flow of time, and symbiotically this geometry dictates how matter is to move. Einstein's elegant theory of gravitation possesses a set of equations which dictate the connection between the matter content of the Universe and its space and time geometry. These are called field equations and they generalize the Newtonian field equation of Poisson, which encapsulates Newton's inverse-square law of gravitation. In addition to this structure, there exist equations of motion which give the analogues of straight lines in the curved geometry. These generalize Newton's laws of motion.

One further erosion of time's absolute Newtonian status occurs in Einstein's theory. Einstein's theory was built upon a premise that there are no preferred observers in the Universe, that is, there is no set of observers for whom all the laws of Nature look simpler. The laws of physics must have the same form for all observers no matter what their state of motion. In other words, however your laboratory is moving—whether it is accelerating or rotating with respect to that of your neighbour—you should both find the same laws of physics to hold good. You may each measure observables to have different values, but you will none the less find them to be linked by the same invariant relationships.

In Einstein's world, there is no special class of observers for whom, by virtue of their motion and time-keeping arrangements, the laws of Nature look especially simple. This is not true in Newton's formulation of motion. His famous laws of motion are found to hold only by experimenters moving in laboratories that are in uniform, non-rotating motion with respect to each other and with respect to the most distant stars, which he took to establish a state of absolute rest. Other observers who rotate or accelerate in unusual ways will observe the laws of motion to have a different, more complicated form. In particular, and in violation to Newton's famous first law of motion, they will observe bodies acted upon by no forces to accelerate.

This democracy of observers that Einstein built into the formulation of his general theory of relativity means that there is no preferred cosmic time. Whereas, in his special theory of relativity, there could exist no absolute standard of time—all time measurements are made relative to the state of motion of the observer—in the general theory of relativity, things are different. There are many absolute times in general relativity. In fact, there appears to be an infinite number of possible candidates. For instance, observers around the Universe could use the local mean density or expansion rate of the Universe to coordinate their time-keeping. Unfortunately, none of these absolute times has yet been found to possess a more fundamental status than the others.

A good way to view an entire universe of space and time (a 'space-time') in Einstein's theory is as a stack of spaces (imagine there to be only two dimensions of space rather than three for the sake of visualization), with each slice in the stack representing the whole universe of space at a different time. The

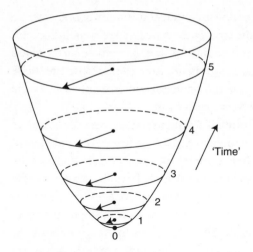

Figure 3.4 Each of the slices 1, 2, 3, 4, and 5 taken through space can be given a 'time' label that is calibrated by the radius of the circle (arrowed). As we progress up the curved surface, the increase in time is gauged by the increasing radii of the circles bounding the slices.

time is a label identifying each slice in the stack. The discussion of the previous paragraph means that we can actually slice up the whole space-time block into a stack of 'time-slices' in many different ways. We could slice through the solid stack at a variety of different angles. This is why it is always more appropriate to talk about space-time rather than the somewhat ambiguous partners space and time. But the connection between matter and space-time geometry means that 'time' can be defined *internally* by some geometrical property, like the curvature, of each slice and hence in terms of the gravitational field of the matter on the slice which has distorted it from flatness (see Figure 3.4 for a simple illustration). Thus we begin to see a glimmer of a possibility of associating time, including its beginning and its end, with some property of the contents of the Universe and the laws which govern how they change.

The new picture of *space-time* rather than space *and* time considerably changes our attitude towards initial conditions and the possible beginning of the Universe. Because of the coupling that exists between the fabric of space-time and matter, any singularity in the material content of space-time (for example, the infinity in the density of matter which occurs in the traditional picture of the Big Bang) signals that space-time has come to an end as well. We now have singularities *of* space and time not merely singularities *in* space and time. Moreover, any space-time given by Einstein's theory of general relativity is an entire Universe. Unlike in Newton's theory, it can never merely describe some object sitting on an external stage of fixed space. Thus the singularities of general relativity are features of the entire Universe, not just one place in it or one moment of its history. These singularities mark out the boundary of space and time.

If we study the expanding Universe according to this picture and trace its history backwards, then it is possible for it to begin at such a singularity. This prediction has been seized upon by many as proof that the Universe had a beginning in time. However, like any logical deduction, this conclusion follows from certain assumptions whose truth needs to be closely examined. The most shaky of these assumptions is that gravity is always attractive. Our modern theories of elementary particles contain many types of particle, and forms of matter, for which this assumption is not true. Indeed, the whole inflationary-universe picture which we introduced above is founded upon the requirement that it be *not* true, for only then can the brief period of accelerated 'inflationary' expansion arise. However, although the avoidance of a singularity might avoid a beginning to time, it would not save us from having to prescribe 'initial' conditions at some past moment to select our actual Universe from the infinity of other possible worlds that begin at singularities. Even if there did exist a singularity, one must face the fact that there are different types of singularity. The specification of the properties of this singularity is an 'initial' condition to be specified on the boundary of our space and time. Some extra ingredient still needs to be found which could provide that specification.

HOW FAR IS FAR ENOUGH?

There was a Door to which I found no key
There was a Veil past which I could not see.
—THE RUBAIYAT OF OMAR KHAYYAM

General relativity (and any other relativistic theory of gravity which does not possess absolutely fixed space or time) gives rise to another subtle property not present in simple Newtonian conceptions of space and time. There are actually many distinct space-times that can arise from the same initial conditions.

Suppose that some space-time S has initial conditions set at some starting time zero which we shall label t_0. We can construct another space-time by removing all of that part of the first space-time that lies to the future of some time t_1 (later than t_0) as well as the time t_1 itself. The new space-time S' is the same as S to the past of the moment t_1, but contains no space or time whatsoever to the future of t_1, as illustrated in Figure 3.5. But both S and S' arise from the same initial state, and indeed we could have cut pieces off S in an infinite number of different ways to make other space-times which

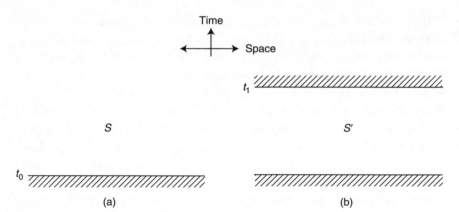

Figure 3.5 Two space-times S and S' which are determined by identical sets of initial conditions prescribed on the surface of initial time t_0. In the case (a), the space-time S is maximally extended, whereas in (b) it is brought to an end arbitrarily at some future time t_1, but no physical infinity or other defect in the space-time structure arises then; the space-time S' is therefore identical to S up to the time t_1, but does not exist to the future of that moment. In practice, it is always assumed that a given set of initial conditions leads to the maximally extended space-time and not one of the infinite number of artificial alternatives which are identical up to some finite moment and then cease to exist for no physical reason.

start from the same initial conditions. Yet there is something unsavoury about S' and its fellow neutered universes. It comes to an end at the allotted time t_1 for no physical reason whatsoever. There is no singularity of any physical quantity. Indeed, we have not had to make mention of the material contents of the Universe at all. The equations that govern the behaviour of matter would still like to predict the future beyond t_1 if only you would allow there to be a future.

This arbitrary truncation of the future is regarded as unrealistically artificial, and cosmologists choose to exclude its possibility and specify the future evolution uniquely. To do so, it is necessary to introduce a further condition into the prescription of possible space-times, or universes, in theories like general relativity, in addition to the specification of initial conditions and laws of Nature. One requires that the Universe should continue to exist until the laws of Nature governing the behaviour of mass and energy signal that time itself has come to an end at a real physical singularity. Under reasonable conditions, it transpires that there is a unique 'biggest' space-time which contains all the others starting from the same initial conditions and which is obtained by letting time go forward until the equations predict a singularity. This *maximally extended* universe is the natural candidate for the space-time that actually arises from a particular set of initial conditions, although we should remember that in principle any of the other truncated realities could be the one that exists following the initial conditions of our Universe. If the maximally extended universe is not the extant one, then the end of the Universe of space and time could indeed come at any moment 'like a thief in the night', without any observable cause or warning.

Despite all these subtleties regarding the nature of time, general relativity has failed to remove the traditional divide between laws and initial conditions. There is still always an initial slice to our space-time stack which determines what the others will look like to its future.

THE QUANTUM MYSTERY OF TIME

It was a book to kill time for those who like it better dead.
— ROSE MACAULAY

In quantum theory, the status of time is an even bigger mystery than it appeared to Newton and Einstein. If it exists in a transcendent way then it is not one of those quantities subject to the famous Uncertainty Principle of Heisenberg, but if it is defined operationally by other intrinsic aspects of a physical system then it does suffer indirectly from the restrictions imposed by quantum uncertainty. Accordingly, when one attempts to produce a quantum description of the entire Universe, one might anticipate some unusual consequences for time. The most unusual has been the claim that a quantum cosmology permits us to interpret it as a description of a universe which has been created from nothing.

The non-quantum cosmological models of general relativity may begin at a definite past moment of time defined using certain types of clock. The initial conditions, which dictate the whole future behaviour of that universe, must be prescribed at that singularity. But, in quantum cosmology, the notion of time does not appear explicitly. Time is a construct of the matter fields and their configurations. Since we have equations which tell us something about how those configurations change as we look from one slice of space to another, it would be superfluous to have a 'time' as well. This is not altogether different to the way in which a pendulum clock tells you time. The clock hands merely keep a record of how many swings the pendulum makes. There is no need to mention anything called 'time'. Likewise, in the cosmological setting, we are labelling the slices in our 'space-time' stack by the matter configuration which creates the intrinsic geometry of each slice. This information about the geometry and material configuration is only available to us probabilistically in quantum theory and it is coded into something which has become known as the *wave function of the Universe*, which we shall henceforth call W.

The generalization of Einstein's equations to include quantum theory is one of the great problems of modern physics. One proposed route uses an equation first found by the American physicists John A. Wheeler and Bryce De Witt. The Wheeler–De Witt equation describes the evolution of W. It is an adaptation of Schrödinger's famous equation governing the wave function of ordinary quantum mechanics but with the curved space attributes of general relativity incorporated as well. If we knew the present form of W, it would tell us the probability that the observed universe would be found to possess certain large-scale features. It is hoped that these probabilities will turn out to be strongly concentrated around particular values in the same way that large everyday things have definite properties despite the microscopic uncertainties of quantum mechanics. If the greatly favoured values were similar to the values observed, then this would give an explanation of those features as a consequence of the fact that ours was one of the most 'probable' of all possible universes. However, to do this, one still requires some initial conditions for the Wheeler–De Witt equation—an initial form for the wave function of the Universe.

The most useful quantity involved in the manipulation and study of W is the transition function $T[x_1, t_1; x_2, t_2]$. This gives the probability of finding the Universe in a state labelled by x_2 at a time t_2 if it was in a state x_1 at an earlier time t_1, where the 'times' can be prescribed by some other attribute of the state of the Universe, for example its average density (see Figure 3.6).

Of course, in non-quantum physics, the laws of Nature predict a definite future state will arise from a particular past one and we would not have use for such probabilistic notions. But, in quantum physics, a future state is determined only as an appropriately weighted sum over all the logically possible paths through space and time that the system could have taken. One of these paths might be the unique one that the non-quantum description would follow. We call this the *classical path*. In some situations, where there exists a conventional deterministic situation, its corresponding quantum description has a transition function that is principally determined by the classical path,

Figure 3.6 Some paths for space-times whose boundary consists of two three-dimensional spaces with curvatures g_1 and g_2 where the matter fields are in the configurations m_1 and m_2 respectively.

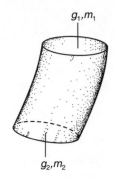

g_1, m_1

g_2, m_2

leaving the others to combine so as to cancel each other out, rather like the peaks and troughs of waves that are out of phase. In fact, it is a deep question whether all possible starting conditions allowed for a quantum universe can give rise to a 'classical' universe when they expand to a large size. This may well turn out to be a very restrictive requirement, one necessary also for the existence of living observers, that marks our Universe out as unusual in the set of all possibilities. If this is true, then it would also have the interesting consequence that only by a study of its cosmological consequences could a complete appreciation of quantum mechanics be arrived at.

In practice, W depends upon the configuration of the matter in the Universe on a particular slice through the space stack and upon some internal geometrical property of the slice (like its curvature) which then effectively labels its 'time' uniquely. Again, there is no special choice of geometrical quantity that is elevated above all others in labelling the slices in this way. There are many that will suffice and the Wheeler–De Witt equation then tells you how the wave function at one value of this internally defined time is related to its form at another value of it. When we are close to the classical path, these developments of the wave function in internal time are straightforward to interpret as small 'quantum corrections' to ordinary classical physics. But this is not always the case, and, when the most probable path is far from the classical one, it becomes increasingly difficult to interpret the quantum evolution as occurring 'in' time in any sense. That is, the collection of space slices that the Wheeler–De Witt equation gives us do not naturally stack to look like a space-time. None the less, the transition functions can still be found. The question of the initial conditions for the wave function now becomes the quantum analogue of the search for initial conditions. The transition function slots x_1 and t_1 are where we could insert our candidates.

QUANTUM INITIAL CONDITIONS

There is no more common error than to assume that, because prolonged and accurate mathematical calculations have been made, the application of the result to some fact of nature is absolutely certain.
— A. N. WHITEHEAD

We have seen that the transition function T tells us about the transition from one configuration of spatial geometry on which the matter has a particular arrangement to another. Let us think of it as $T[m_1, g_1; m_2, g_2]$, where m labels the matter configuration and g is some geometrical characteristic of space, like the curvature, which we are using as an internally defined time at two values '1' and '2'. We can envisage universes that begin at a single point rather than at an initial space, so that their development looks conical rather than cylindrical (as was the case in Figure 3.6). This is illustrated schematically in Figure 3.7. Yet this is no great advance in our attempt to transmogrify the idea of initial conditions, because the singularity of the non-quantum cosmological models always shows up as a feature of the classical quantum path, and in any case we just seem to be picking a particular initial condition, which happens to describe creation from an initial pre-existent point, for no good reason. We have not severed the dualism between laws (represented here by the Wheeler-De Witt equation) and initial conditions.

There is a radical path that may now be taken. One should stress that it may well turn out to be empty of any physical significance. It is an article of faith. If we look at Figures 3.6 and 3.7, then we can see how the stipulation of an initial condition g_1 relates to the state of the space further up the tube or the cone at g_2. Could the boundaries of the configurations at g_1 and g_2 be combined in some way so that they describe a single smooth space which contains no nasty singularities?

We know of simple possibilities in two dimensions, like the surface of a sphere, which are smooth and free of any singular points. So we might try

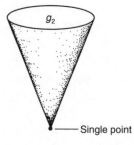

Figure 3.7 A space-time path whose boundary consists of a curved three-dimensional space of curvature g_2 and a single initial point, rather than another three-dimensional space. If there is a singularity in the curvature or the matter configuration at the point, we cannot calculate the transition probability T from this point to the state with curvature g_2. If this had been possible, it would give the probability of a particular type of universe arising from a 'point' rather than from 'nothing'.

Figure 3.8 An appealing path is one whose boundary is smoothly rounded off so that it consists of just a single three-dimensional space, with no conical 'point' at the base as there was in Figure 3.7. This admits an interpretation of the transition probability as creation out of 'nothing' because no initial state exists: there is a single boundary. This can be employed as the picture of the three-dimensional boundary of a four-dimensional space-time only if we suppose that time behaves like another dimension of space.

to conceive of the whole boundary of the four-dimensional space-time to be not g_1 *and* g_2 but a single smooth surface in three dimensions. This might be the surface of a sphere sitting in four space dimensions. One of the curious and attractive features of these smooth surfaces that mathematicians habitually consider regardless of their dimension, which we can visualize better by returning to the two-dimensional surface of an ordinary sphere, is that they are finite in size but nevertheless have no edge: the surface of the sphere has a finite area (it would only require a finite amount of paint to paint it), but however one moves one never runs into an unusual point like the apex of a cone. We might describe the sphere as being without boundary from the point of view of flatlanders living on its surface. Interestingly, such a configuration can be conceived for the initial state of the Universe (see Figure 3.8). However— and now comes the radical step—the sphere we are using as an example is a space of three dimensions with a two-dimensional surface as a bondary. But, for our quantum boundary, we need a three-dimensional space as a boundary. However, this requires the four-dimensional thing of which it is the boundary to be a four-dimensional *space* and not a four-dimensional *space-time*, which is what the real Universe has always been assumed to be. Therefore, it is proposed that our ordinary concept of time is transcended in this quantum cosmological setting and becomes like another dimension of space, so making three-plus-one dimensions of space and time into a four-dimensional space. This is not quite as mystical as it might sound because physicists have often carried out this 'change time into space' procedure as a useful trick for doing certain problems in ordinary quantum mechanics, although they did not imagine that time was *really* like space. At the end of the calculation, they just swop back into the usual interpretation of there being one dimension of time and three other qualitatively different dimensions of what we call space.

The radical character of this approach is that it regards time as being truly like space in the ultimate quantum gravitational environment of the Big Bang. As one moves far away from the beginning of the Universe, so the quantum effects start to interfere in a destructive fashion and the Universe is expected to follow the classical path with greater and greater accuracy. When this happens, the conventional notion of time as a distinct concept to that of space begins to crystallize out. Conversely, as one approaches the beginning, so the conventional picture of time melts away and time becomes indistinguishable from space as the effects of the boundary condition are felt.

This 'no boundary' condition was proposed by James Hartle and Stephen Hawking for aesthetic reasons. It avoids singularities from the initial state and removes the conventional dualism between laws and initial conditions. This it can achieve if the distinction between space and time is lost. More precisely, the 'no boundary' proposal stipulates that, in order to work out the wave function of the Universe, we compute it as the weighted aggregate of paths which are restricted to those four-dimensional spaces which possess a single finite smooth boundary like the spherical one we have just discussed. The transition probability that this prescription provides for the production of a wave function with some other matter content m_2 in a geometrical configuration g_2 just has the form $T[m_2, g_2]$. Thus there are no slots corresponding to any 'initial' state characterized by m_1 and g_1. Hence, this is often described as giving a picture of 'creation out of nothing', in which T gives the probability of a certain type of universe having been created out of nothing. The effect of the 'time becomes space' proposal is that there is no definite moment or point of creation. In more conventional quantum mechanical terms, we would say that the Universe is the result of a quantum mechanical tunnelling process, where it must be interpreted as having tunnelled from nothing at all. Quantum tunnelling processes, which are familiar to physicists and routinely observed, correspond to transitions which do not have a classical path.

THE GREAT DIVIDE

I sometimes ask myself how it came about that I was the one to develop the theory of relativity. The reason, I think, is that a normal adult never stops to think about problems of space and time. These are things which he has thought of as a child. But my intellectual development was retarded, as a result of which I began to wonder about space and time only when I had already grown up.
— ALBERT EINSTEIN

The overall picture one gets of this type of quantum beginning is that the Wheeler–De Witt equation gives the law of Nature which describes how the wave function W changes. The geometry of the space can be used as a measure of time which looks essentially like the ordinary time of general relativity when one is far from the Big Bang. But, as one looks back towards that instant which we would have called the zero of time, the notion of time fades away and ultimately ceases to exist. This type of quantum universe has not always existed; it comes into being just as the classical cosmologies could, but it does not start at a Big Bang where physical quantities are infinite and where further initial conditions need to be specified. In neither case is there any information as to what it may have come into being from.

We should stress again that this is a radical proposal (which we shall discover, in Chapter 5, can become even more radical). It has two ingredients: the first is the 'time becomes space' proposal; the second is the addition of the 'no boundary' proposal—a single prescription for the state of the Universe, which subsumes the roles of both initial equations and laws of Nature in the traditional picture. Even if one subscribes to the first ingredient, there are many choices one could have used instead of the second to specify the state of a Universe which tunnels into existence out of nothing. These would all have required some additional specification of information.

The study of the wave function of the Universe is in its infancy. It will undoubtedly change in many ways before it is done. The 'no boundary' condition leaves much to be desired. It probably contains too little information to describe all the observable features of a real universe containing irregularities like galaxies. It must be supplemented by additional information about the matter fields in the Universe and how they distribute themselves. Of course, it may also be complete nonsense. The important lesson for us to draw from it here is the extent to which our traditional dualism regarding initial conditions and laws might be mistaken. It might be an artefact of our experience of a realm of Nature in which quantum effects are small. If a theory of Nature is truly unified, then we might expect that it would exploit the possibility of keeping time in terms of the material contents of the Universe so as to marry together the constituents of Nature with the laws governing their change and the nature of time itself. However, we are still left with a choice as to the boundary condition which should be imposed upon some entity like the wave function of the Universe. No matter how economical its prescription, it is an inescapable fact that the 'no boundary' condition and its various rivals are picked out only for aesthetic reasons. They are not demanded by the internal logical consistency of the quantum universe.

The dualistic view that initial conditions are independent of laws of Nature must be reassessed in the case of the initial conditions for the Universe as a whole. If the Universe is unique—the only logically consistent possibility—then the initial conditions are unique and become in effect a law of Nature themselves. This is the motivation of those who seek basic principles which might serve to delineate the initial conditions of the Universe. If this is truly the case, then it introduces another new ingredient into our thinking about the Universe, because it points to a fundamental asymmetry between the past and the future in the make-up of the laws of Nature. On the other hand, if we believe that there are many possible universes—indeed may actually *be* many possible universes 'somewhere'—then initial conditions need have no special status. They could be just as in more mundane physical problems: those defining characteristics that specify one particular actuality from a general class of possibilities.

The traditional view that initial conditions are for the theologians and evolution equations for the physicists seems to have been overthrown—at least temporarily. Cosmologists now engage in the study of initial conditions to discover whether there exists a 'law' of initial conditions, of which the 'no boundary' proposal would be just one possible example. This is radical indeed, but perhaps it is not radical enough. It is worrying that so many of the concepts and ideas being used in the modern mathematical description— 'creation out of nothing', 'time coming into being with the Universe'—are just refined images of rather traditional human intuitions and categories of thought. Surely, it is these traditional notions that motivate many of the concepts that are searched for and even found within modern theories that are cast in mathematical form. The 'time becomes space' proposal is the one truly radical element that we cannot attribute to our inheritance of past generations of human thinking in philosophical theology. One suspects that a good many more habitual concepts may need to be transformed before the true picture begins to emerge.

CHAPTER 4

Forces and particles

A vacuum is a hell of a lot better than some of the stuff that nature replaces it with.
—TENNESSEE WILLIAMS

THE STUFF OF THE UNIVERSE

The scenery in the play was beautiful, but the actors got in front of it.
— ALEXANDER WOOLLCOTT

Devices of all sorts, whether they be computers or milling machines, need to act upon suitable raw material. If you design a spanner on general mechanical principles with a little bit of symmetry thrown in for appearances, it will none the less be useless if it fails to fit your particular bolt-head shapes. Likewise, a Theory of Everything needs information about what particles and forces actually exist. A knowledge of the laws of Nature is of little use unless one knows what it is that those laws govern. In this respect the contrast between the traditional classical physics of Newton and the elementary-particle world is striking. Newton emphasized the universality of his laws of motion: they apply without exception to 'all bodies' irrespective of their other idiosyncrasies. Yet it is this very universality that prevents the laws of classical physics having anything to say about what particles or bodies actually do exist. They focus upon certain universal attributes of particles, like their mass, to the exclusion of all others. To those of us who have grown up learning about Newtonian mechanics from schooldays, this seems a familiar and reasonable approach but how difficult it must have been for the first students of motion to identify the salient features of a real object which should be included in the laws of motion.

A beautiful illustration of the dilemma is presented by the French scientist Moreau de Maupertuis, the originator of the Principle of Least Action during the eighteenth century. With regard to the laws of momentum conservation which govern the collision between bodies like snooker or pool balls, he observes:

> If someone who had never touched a body or seen how bodies collide, but who was experienced in mixing colours, saw a blue object move toward a yellow one, and were asked what would happen if these two bodies collide, he would probably say that the blue body would turn green as soon as it united with the yellow one.

It is not difficult to appreciate why the number of properties of a body which can be important for its dynamics has to be minimal. Typical good-sized objects, like rocks, footballs, or cars have so many individual properties that if the laws governing their motion were closely associated with many of their defining properties then it would be as good as having no laws at all. Every rock, car, or billiard ball is different in a myriad of ways and each would respond to the same law in very different ways. Such a situation is very similar to that found in many early Greek writings. They did not readily have the notion of an external Lawgiver in Nature who dictated external laws of Nature. Instead, they were partial to the notion that bodies contained immanent tendencies which dictated how they would move. Whereas Plato sought to understand what was observed in the world in terms of another world of perfect blueprints of which observed things were but an imperfect approximation, Aristotle believed that these 'forms' which dictated how things behaved were not inhabitants of some abstract other-world but were *in* the things themselves. Aristotle's ideas held sway for thousands of years, until they were discarded because of a combination of religious and scientific considerations. Newton dismissed this tradition of innate tendencies rather forcefully in his correspondence with Richard Bentley:

> That gravity should be innate, inherent and essential to Matter...is to me so great an Absurdity, that I believe that no Man who has in philosophical matters a competent Faculty of thinking, can ever fall into it.

Newton saw that it was necessary to discard this view if one was to move forward and separate what we do know from what we do not. No universal laws could emerge if we regarded laws of Nature as innate to the particles they governed. The future course of physics until the early twentieth century therefore regarded the material content of the Universe as logically distinct from the laws that governed it. The former had to be discovered by observation whilst the laws which governed the behaviour of particular things acted upon a very small number of attributes, like electric charge or mass, whose identities were revealed by our accumulated experience.

THE COPY-CAT PRINCIPLE

Repetition is the only form of permanence that nature can achieve.
— GEORGE SANTAYANA

The laws of elementary-particle behaviour are different precisely because the objects they govern are not different. Whereas the rocks and billiard balls of classical physics are all different the most elementary particles of matter fall into classes of *identical* particles: all electrons are the same, all muons the same, and so on, throughout the elementary-particle world. It is a world of clones. Once you have seen one electron, you have seen them all. But it is this copy-cat principle that makes it possible for the laws which govern the behaviour of electrons and muons to be closely linked to the intrinsic properties of electrons and muons without sacrificing their universality. It also plays a crucial role in our human quest to understand the Universe, for it underpins our belief that by an exhaustive study of a small part of the Universe we can approach an understanding of the whole.

The fact that Nature displays populations of identical elementary particles is its most remarkable property. It is the 'fine tuning' that surpasses all others. Our experience of the Universe has never given us any reason to doubt the assumption that all electrons are the same, all photons are the same wher- ever and whenever they are. In the nineteenth century, James Clerk Maxwell highlighted the fact that the physical world was composed of identical atoms which were not subject to gradual mutation or evolution. Today, we look for some deeper explanation of the elementary particles of Nature in terms of a Theory of Everything like string theory. One of the perplexing features of the successful theories of the electromagnetic, weak and strong interactions of particle physics that provoked particle physicists to search for a deeper under- lying theory was the profusion of elementary particles. There were so many of them that it suggested that there was a smaller and more basic population of entities inside them. Perhaps they were not elementary at all? Could they be composed of different combinations of a far smaller number of elementary objects? Attempts to make theories of possible building blocks were never very compelling and led to no testable predictions.

String theories offer another route to solving this problem. Instead of a Theory of Everything containing a population of elementary point-like particles, string theories introduce basis entities that are lines or loops of energy which have a tension. As the temperature rises the loops shudder and vibrate in an increasingly stringy fashion, but as the temperature falls the tension increases and the loops become more and more point-like. So, at low energies the strings behave like points and allow the theory to make the same types of accurate prediction about what we should see as the intrinsically point-like theories do. However, at high energies, things are different. The hope is that it will be possible to determine the principal energies of vibration of the superstrings. All strings, even guitar strings, have a collection of special vibrational energies that they naturally take up when disturbed. It is hoped that, if we could calculate these special energies for the superstring, then they would (by virtue of Einstein's famous formula of mass–energy equivalence: $E = mc^2$) correspond to a collection of masses that correspond in some way to the 'particles' that we call elementary. So far, these energies have proved too hard to calculate. However, one of them has been found: it corresponds to a particle with zero mass and two quantum units of spin. This spin ensures that it mediates attractions between all masses. It is the particle that we call the 'graviton' and shows that string theory necessarily includes the phenomenon of gravitation—a remarkable and compelling feature since earlier candidates for a Theory of Everything all failed miserably whenever they were challenged to find a way to include gravity in the unification story.

It is this repeatability of things that is the hallmark of the most basic entities in Nature and at root it is the reason why there can be accuracy and reliability in the physical world, whether it be in DNA replication or in the stability of the properties of matter. But we shall find it opens up the possibility that the strict divide between the laws of Nature and the entities that they govern may be compromised when we probe Nature more deeply, just as in the last chapter we found it possible to muddy the divide between laws and initial conditions in the quantum description of the Universe. If there exists a real divide between the constituents of the Universe and the laws that govern them, then any Theory of Everything would require additional information to restrict the identities of particles. This seems unsatisfactory to amateur universe builders like ourselves. We would expect that things could be perfectly unified in some sense, so that the laws and the ultimate particles of Nature that they govern are married together in a union of perfect and unique intercompatibility. The laws should decree what their subjects are in addition to what they may do.

This symbiosis of laws and particles and forces has begun to come to pass in modern physics as a result of the discovery of a breed of physical theory called a *gauge theory*. All the best theories of the fundamental forces of Nature—of gravity, of electromagnetism, and of the weak and strong nuclear forces—are gauge theories. Let us dwell upon how this threefold union of particles, forces, and laws comes about within this jurisdiction.

For the Newtonian physicist whose laws governed the behaviour of objects on an absolute space moving through unbending time, forces moved things in a mysterious way. Gravity acted instantaneously between masses by a process that Newton found it fruitless to enquire into any further. Gradually, throughout the twentieth century, the effect of the cosmic speed limit for the transfer of information imposed by Einstein's special theory of relativity has made its presence felt. Instantaneous gravitational effects would violate that limit by allowing signals to be transmitted faster than the speed of light in a vacuum. As a result, we picture forces of Nature as mediated by the exchange of particles between the bodies which are in interaction. Thus the gravitational force is mediated by the exchange of gravitons, the electromagnetic force by the exchange of photons, the weak interaction by the exchange of massive W or Z particles, and the strong interaction between quarks by the exchange of gluons. In some cases, these exchange particles actually feel the force which they mediate. This is the case for gravitation and for the strong and weak interactions, although not for the electromagnetic interaction which acts between electrically charged elementary particles. The interactions between such particles are mediated by the exchange of a photon of light which is uncharged. Thus we see that the forces of Nature are deeply entwined with the elementary particles of Nature. They cannot be considered independently.

The other arms of the golden triangle, the connection between forces and particles and the laws themselves exist only in these elegant creations called gauge theories. Their emergence has undermined a longstanding prejudice regarding the Galilean and Newtonian revolutions in the description of Nature that scientists ceased asking 'why' questions of Nature and were content to know only 'how' things were. Curiously, modern particle physicists are quite different. Gauge theories show that physicists need not be content to possess theories that are perfectly accurate in their description of *how* particles move and interact. They can know something of *why* those particles exist and *why* they interact in the manner seen.

The most successful fundamental theories of physics—general relativity (the theory of gravity), quantum chromodynamics (the theory of the strong sub-nuclear forces between quarks and gluons), and the Weinberg–Salam theory (the unified theory of the electromagnetic and weak interactions)—are all theories of a particular type known as *local gauge theories*. We have already seen in Chapter 2 how certain geometrical invariances of the laws of Nature are equivalent to the imposition of physical invariances. For each symmetry, there exists an associated conserved quantity. This correspondence is maintained even when the symmetries involved are more esoteric than simple rotations or translations in space. These additional invariances are called internal symmetries and correspond to invariances under various relabellings of the particles involved, for example swapping the identities of all the protons and neutrons in the Universe. Gauge symmetries are different again. They do not lead to conserved quantities in Nature; rather, they impose powerful requirements upon the form and scope of the laws of Nature. In particular, they dictate what forces of Nature exist and the properties of the elementary particles which they govern. The simplest example is that of a *global gauge symmetry*. It demands that the world be invariant if we shift every point in the same way. Imagine such an operation performed upon an object like your hand. It would be transported in space but would look the same. But it is unnatural to suppose that the changes be the same everywhere. If a particle changes at this moment on the other side of the Universe, then a particle here and now cannot know this at least until a light signal has had time to pass between them. It would require instantaneous signalling to keep in step. Global gauge invariance is a somewhat unappealing restriction that retains echoes of Newton's instantaneous action at a distance. This leads us to demand the more realistic, but much more stringent, requirement that things be invariant under *local gauge symmetry*, wherein every point can change in a different way.

Invariance in this case seems impossible. In our earlier example, every part of your hand would move off in different directions. The only way in which things can be kept invariant under such general changes is if certain forces exist which constrain the allowed motions. Imagine some elastic bands taut around your hand which restrict the ways in which parts of it can move: the elementary-particle world is akin to having an infinite network of entwined constraints like this which transform all possible changes into a small class of particular ones. In this way, the imposition of invariance under local gauge symmetry actually dictates what forces of Nature exist between the particles involved. They reveal why there must be electromagnetism as well as how it operates.

Einstein's general theory of relativity is a local gauge theory of this sort. Einstein wished to generalize the Principle of Special Relativity which maintained that the laws of physics be the same for all observers moving at constant relative velocities to the situation where they accelerate. The only way in which this is possible for all observers in arbitrary accelerated motion is for there to exist a gravitational field.

The Platonic faith in symmetry and the implementation of those symmetries as the basis for gauge theories is the foundation of our knowledge of elementary-particle interactions. Yet is does not tell us everything. It fails to tell us how many particles of a similar type there must be. Why are there three types of neutrino rather than just one. Why there is only one variety of photon. Nature appears to have used a 'copy-cat' principle in two ways. It has created populations of identical particles like electrons and electron-type neutrinos; but it has also created muons and muon-type neutrinos and tau particles and their associated neutrinos. These are similar to the electron and its neutrino in many ways. What one would like to know is why there exist these small variations on the same major theme and why there are just three of them and no more. The different gauge theories have failed to tell us how many of these copies there must be. Moreover, to complete a fully unified picture of the Universe, we must do something about the fact that we have many different gauge theories which must be unified into a single description by embedding the different symmetries associated with individual theories into a bigger over-riding pattern, or *grand unified theory*. Grand unification removes the problems of different disjoint theories, but it still does not solve the problem of what limits the number of types of similar particles.

Gauge theories, by their very nature, are built upon symmetry. These symmetries are built up by operation of a finite number of variations upon a single theme. There are only a finite number of basic generators of the possible patterns that span all the possibilities compatible with the maintenance of a particular symmetry. The greater the number of basic generators so the larger the range of patterns. Furthermore, the basic generators of the set of patterns consistent with any underlying symmetry define that symmetry and correspond to the 'carrier' particles which mediate the forces of Nature. Thus, in Maxwell's theory of electromagnetism, there is just one generator of the symmetry and this corresponds to the photon; the symmetry governing the weak force has three generators corresponding to the electrically neutral Z boson and the positively and negatively charged W bosons; the strong force between quarks has eight generators corresponding to the eight varieties of gluon carrying the three varieties of the type of charge called 'colour' which the strong force recognizes. As a result, we see that there is a certain element of *finiteness* built into all of these theories. The finiteness of the symmetry is associated with the finite number of the elementary particles that are the basic generators of the symmetry. A world which had a bottomless infinity of elementary particles would be well on the way to anarchy. Its symmetries would need to be so large that their influences would be tantalizingly weak.

ELEMENTARITY

We spoke of the 'Properties of Things', and of the degree to which these properties could be investigated. As an extreme thought, the following question was proposed: Supposing it were possible to discover all the properties of a *grain of sand*, would we then have gained a complete knowledge of the *whole universe?* Would there then remain no unsolved component of our comprehension of the universe?
—A. MOSZKOWSKI

The most topical aspect of the identification of the forces and particles of Nature is to know the identity of the most elementary entities in Nature. Until only a few years ago, they were invariably imagined to be idealized 'points' of zero size. Quarks and leptons were taken to be particles of this sort, exhibiting no evidence of internal structure in any particle scattering experiment. If a particle physicist were asked how many angels can dance on a quark, he could answer none without a moment's hesitation. However, theories in which the most basic entities are points—quantum field theories as they are known—possess unpleasant mathematical properties. They lead to mathematical infinities that must be ignored in the process of calculating observable quantities. This can usually be done by following a systematic recipe which amounts to ignoring the infinite part of any answer, but the procedure is aesthetically rather unappealing. It has only been tolerated in practice because the finite parts that remain in these calculations after the infinite parts have been removed produce predictions of observed quantities that are correct to fantastic precision. There is clearly a deep truth somewhere close to the heart of this picture.

It has now been recognized that theories in which the most elementary objects are lines or loops ('strings'), rather than points, can avoid these defects. Moreover, whereas the point particle schemes require a separate point endowed with characteristics like mass to be specified for each elementary particle separately, a single string possesses an infinite number of modes of vibrations, just like the harmonics of a violin string, and the energy of each different mode will correspond to a different elementary-particle mass (via the mass–energy equivalence $E = mc^2$). Most of this collection of particle masses will be concentrated around unobservably high energies, but the others should include the masses of the known elementary particles. Furthermore, the number of copies of each type of particle appears to be tied to the underlying symmetrical structure of these theories. They have the scope to tell us why there are three varieties of neutrino at low energy. Whereas earlier elementary-particle theories could provide no explanation for this aspect of things, string theories link it to the laws of Nature in a deep way and turn it into an answerable 'why' question. This explanatory potential is the great hope of string theories and is the hallmark of their claim to be a Theory of Everything. They should contain within them the deep connection between the symmetries or laws of Nature and the entities which those laws govern, but as yet the difficulty of extracting that information from the theory has proved insurmountable. It is one thing to have the Theory of Everything; quite another to solve it. One day it is hoped that definite predictions of the masses of the elementary particles of Nature will be extracted from this theory and compared with observation.

Strings aim to explain all the properties of the elementary particles of Nature. But in terms of what will they explain them? What are the properties of strings themselves? Strings possess one defining property which is their tension. This quantity plays a crucial role in the overall picture of how strings can be reconciled with the miraculous experimental success of the point-like quantum field theories in explaining the observed features of the world at lower energies. For the strings possess a tension that varies with the energy of the environment, so that at low energies the tension is high and pulls the strings taut into points and we recover the favourable features of a world of point-like elementary particles. At high energies, where the string tension is low, their essential stringiness becomes evident and creates behaviour that is qualitatively different from that of the point-particle theories. Unfortunately, at present, the mathematical expertise required to reveal these properties is somewhat beyond us. For the first time, modern physicists have found that off-the-shelf mathematics is insufficient to extract the physical content of their theories. But, in time, suitable techniques will no doubt emerge, or perhaps a better way to look at the theory will be found: one that is conceptually and technically simpler.

In summary, we have seen that we need to know the identity of the forces and particles of Nature. At present, we believe, perhaps mistakenly, that we have identified all the fundamental forces. We have working gauge theories, based upon particular group symmetries which determine the structure of these forces and actually tell us why they must exist if certain symmetries are to be maintained by the laws of Nature. Schemes exist which unify these different gauge theories together, but they fail to limit the number of types of particle that can exist. Ultimately, the demand for self-consistency alone narrows the range of options for the single over-arching symmetry of Nature from which everything else follows. Yet this route requires a more radical ingredient for its successful implementation. This has led to the abandonment of our belief that the most basic entities in Nature are points. The string theories that have emerged as a result of this desire to unify without flaws are narrowly pinned down to possess only a small number of possible over-arching symmetries.

In the march towards such a self-consistent single description of the forces of Nature, the traditional viewpoint that the most basic theories of physics must be quantum field theories has been undermined by the theoretical attractions of string theories and their promise to explain the properties of all the elementary particles of Nature. At present, strings are *all* theory. In the future, we hope that their multitude of properties can be extracted. Nevertheless, any theory based upon symmetry always has the spectre of a larger symmetry hanging over it. How do we know that our entire scheme, no matter how internally consistent and experimentally successful it may ultimately prove to be, does not lie within some far bigger scheme of things defined by consistency with respect to properties of the Universe which we have yet to envisage, associated with feeble forces of Nature we have yet to witness directly?

THE ATOM AND THE VORTEX

Anticipatory plagiarism occurs when someone steals your original idea and publishes it a hundred years before you were born.
— ROBERT MERTON

The introduction of the 'string' as the basis for explaining the nature of elementary particles and their interactions is an example of the deployment of topology in physics. Topology is that branch of mathematics which is interested in the forms of things aside from their size and shape. Two things are said to be topologically equivalent if one can be deformed smoothly into the other without sticking, cutting, or puncturing it in any way. Thus an egg is equivalent to a sphere. The first application of topology to an analogous problem—the interaction of atoms rather than elementary particles—was made in the mid-nineteenth century by Lord Kelvin. It has many striking parallels with the aims and attractions of modern string theory.

In 1867, Kelvin presented a new theory of atoms to the Royal Society of Edinburgh, and subsequently published a written version of his ideas in the Society's journal. He had been much impressed by Helmholtz's investigations into the behaviour of interacting vortices in liquids, which he had seen demonstrated by his friend Tait in a series of ingenious experiments with smoke-rings. Kelvin wished to view atoms as some form of local eddy in the state of a universal fluid that permeated the Universe. Helmholtz had shown that vortex filaments in a perfect liquid could remain in a stable state immune from dissipation. Vortices thus meet one of the necessary requirements of any theory of matter. Of Tait's demonstrations, he writes:

> A magnificent display of smoke-rings, which he recently had the pleasure of witnessing in Professor Tait's lecture room, diminished by one the number of assumptions required to explain the properties of matter on the hypothesis that all bodies are composed of vortex atoms in a perfect homogeneous liquid. Two smoke-rings were frequently seen to bound obliquely from one another, shaking violently from the effects of the shock. The result was very similar to that observable in two large india-rubber rings striking one another in the air. The elasticity of each smoke-ring seemed no further from perfection than might be expected in a solid india-rubber ring of the same shape, from what we know of the viscosity of india-rubber. Of course this kinetic elasticity of form is perfect elasticity for vortex rings in a perfect liquid. It is at least as good a beginning as the 'clash of atoms' to account for the elasticity of gases.

Kelvin envisaged a picture of atomic interactions in which each atom was a vortex in some ethereal background fluid. The observed stability of atoms had its parallel in the striking stability of the vortex rings he had observed and which could be traced to Helmholtz's discovery that a measure of the circulation of a vortex system was conserved in any interactions between them. A single vortex could not just be created out of nothing. Vortices could only appear in equal and opposite pairs. He also recognized that it was possible to explain an enormous variety of atomic structures by exploiting the vast array of different knotted configurations in which vortex tubes could arrange themselves, and he envisaged

> knotted or knitted vortex atoms, the endless variety of which is infinitely more than sufficient to explain the varieties and allotropies of known simple bodies and their mutual affinities.

In effect, the array of all possible knots is available to the vortices. In fact, this provoked Tait to undertake a detailed study of the classification of knots. But the last property of the vortices to which he appealed was the most striking. For it is one of the key features of modern string theory that each string has associated with it the energies of its natural modes of vibrations and these can be associated with the mass-energies of elementary particles. Kelvin hoped to explain the spectral lines of the chemical elements in terms of the natural modes of vibration of the vortices which constituted them. Again, he appeals to the observed stability properties of such vibrations as an admirable basis for such a theory:

> The vortex atom has perfectly definite modes of vibration, depending solely on that motion the existence of which constitutes it. The discovery of these fundamental modes [of vibration] forms an intensely interesting problem for pure mathematics.

These ideas led him on to make further fascinating speculations: that there can exist atomic structures which are formed from chains of interlocking vortices, and that the vibrational energies of the vortices should exhibit a dependence upon temperature which might lead to the phenomenon of absorption by coinciding with the vibrational modes of another substance.

Kelvin and his colleagues worked seriously on this theory for nearly two decades and it was taken seriously by leading physicists of the time. But it was eventually abandoned for the lack of any definite successes. Re-examined in the light of modern string theory, it displays a remarkable early picture of how stability can arise from purely topological changes and how the presence of vibrational modes might be the source of stable energetic configurations of matter.

A WORLD BESIDE ITSELF

'I am half sick of shadows,' said
The Lady of Shalott.
— ALFRED LORD TENNYSON

Behind this picture of a world of stringy things, there lurks the prospect of a yet more radical picture. There may be a good deal more to the Universe than meets the eye, even the eye of faith of the cosmologist. Einstein's theory of gravitation has taught us that the notion of force may be nothing more than a convenient anthropomorphism. The classical picture of physical laws sees them as sets of rules which dictate how particles respond to the action of certain 'forces' between them when the particles are set down in the traditional space whose geometry was laid bare by Euclid. Einstein's general theory of relativity provided us with a picture of gravitation that was altogether more sophisticated. The presence of the particles of matter, and their motion, determine the local topography of the space in which they sit. No longer are there mysterious forces acting between neighbouring bodies. Each now moves along the most economical path available to it on the undulating space created by all the particles in the Universe. Thus the Sun creates a large ditch in the space near the Earth, and the Earth moves around the inside surface of that ditch. This path we call its orbit. There are no gravitational 'forces' acting between distant objects. Everything takes its marching orders from the spatial topography of its immediate locale.

Hence, although students of Einstein's theory of gravitation talk about 'gravitational forces', they do so really just out of habit. The concept of force has been subsumed within the more elegant and powerful conception of a dynamic space-time geometry. We might suspect, therefore, that any candidate for a truly fundamental Theory of Everything, like string theory, that includes and supersedes Einstein's picture of gravitation and unites it with the other forces of Nature may lead to the dissolution of these other forces as well. Perhaps the search for the Theory of Everything will reveal to us that these fundamental forces of Nature, whose unification we have expended so much effort upon, are like the inhabitants of Prospero's enchanted isle

> ...all spirits, and
> Are melted into air, into thin air;
> And, like the baseless fabric of this vision,...
> Leave not a rack behind.

String theory promises to take a further step beyond that taken by Einstein's picture of force subsumed within curved space and time geometry. Indeed, string theory contains Einstein's theory of gravitation within itself. Loops of string behave like the exchange particles of the gravitational forces, or 'gravitons' as they are called in the point-particle picture of things. But it has been argued that it must be possible to extract even the geometry of space and time from the characteristics of the strings and their topological properties. At present, it is not known how to do this and we merely content ourselves with understanding how strings behave when they sit in a background universe of space and time. But the stringy picture of those gravitational forces which we have already wedded so closely to the nature of space and time promises to create a number of new perspectives on things. For example, if the Universe is envisaged to collapse back upon itself in the future and contract towards a state of ever higher density, what will be the end result? The conventional point-particle picture allows the collapse to proceed to a true singularity of infinite density in a finite time. But, in the string picture, we could envisage the energy of collapse being soaked up, exciting all the possible vibrational states of the strings, and the collapse could thus be halted. The strings act like cosmic shock-absorbers. Conversely, perhaps the initial state of the whole Universe corresponds to some unusual string state which releases its internal vibrational energy into expansion energy.

It is the investigation of these very high density regimes where gravity and quantum mechanics both influence events that string theory hopes to make its biggest impact upon our picture of the world by providing us with a theory of quantum gravitation. It has very promising credentials because, whereas other theories of quantum phenomena always develop inconsistencies when any attempt to incorporate gravity into them is made, string theory demands that gravity exist in order to be consistent. To illustrate how a string picture could capture the essence of quantum gravitational phenomena, one can trace the situation pictured in Figure 4.1. Consider a loop of string as it moves through space and time. It traces out the world tube shown in Figure 4.1(a). But quantum fluctuations and uncertainties would cause this tube to have an erratic surface as shown in Figure 4.1(b). Now take a slice through the fluctuating string at particular times, then what you see is shown in Figure 4.1(c). The picture obtained is identical to the presence of a number of interacting loops rather than just the single loop that constituted the non-fluctuating state in Figure 4.1(a). This simple picture illustrates how it is possible to include the quantum gravitational fluctuation effects into a string theory that contains simple loop interactions.

Another important feature of the Universe for the physicist is the fact that there appear to exist four basic forces of Nature from which all natural phenomena flow. There may exist other very weak forces of which we have had no obvious experience, and were this to be the case then the task of producing an all-encompassing theory of them all would be far more difficult than anticipated. The other important feature of the forces of nature is that they are different. They act upon different sub-collections of particles and they possess different strengths. The gravitational force, the strong nuclear force, the electromagnetic force, and the weak force differ in relative strengths roughly as 10^{-39}, 1, 10^{-2}, and 10^{-5}. This wide-ranging spectrum plays an

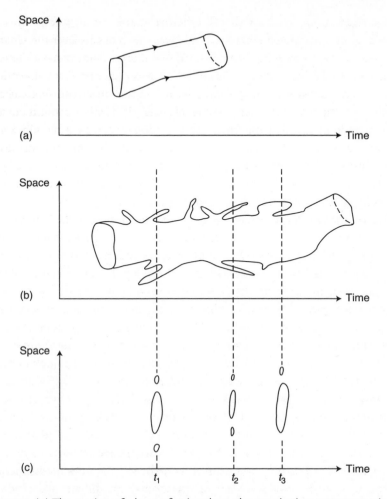

Figure 4.1 (a) The motion of a loop of string through space in time traces out a 'world tube' in space-time; (b) the effects of quantum gravitational fluctuations produce a foam-like distortion of the simple 'world tube' displayed in (a); (c) slices through (b) at three instants of time, t_1, t_2, and t_3. At each instant of time, the effect of the distortions shown in (b) on the basic 'world tube' of (a) are equivalent to interactions between loops of string.

important role in our quest to understand the Universe. If the forces were to act upon all particles with similar strength, then the world would be vastly more complicated. All the forces would be important in almost every situation. The hierarchy of force strengths ensures that this does not happen. The structures that we observe in the Universe are balancing acts between pairs of natural forces with the others playing a negligible role.

Nevertheless, we might well ask whether there *are* only four forces of Nature. More than fifteen years ago, there was much debate over the reality of a so-called 'fifth' force. It has been claimed that there is evidence that Newton's inverse-square law of gravitational attraction is not the true behaviour of the force between masses when gravity is weak. Rather, there is a small change in this law which is equivalent to the addition of another force to it. This extra ingredient is called the 'fifth' force, although strictly it should just be interpreted as the hypothesis of a slightly different behaviour for the known force of gravity. Most physicists are sceptical of the existence of this 'fifth' force and the most recent and accurate experiments have not confirmed the original claims of the first experimenters, who suggested there was evidence for such a force of Nature. Rather than dwell on this dispute, it is more instructive to dwell upon why there can be such a dispute. Gravity is an extraordinarily weak force. On the scale of atoms and everyday objects, it is ten followed by thirty-seven noughts times weaker than the other forces of Nature. Hence it is very difficult to detect. Its effects are overwhelmed by the other forces: magnets stop pieces of metal falling to the ground; the sub-atomic forces of Nature prevent elementary particles just falling into a heap on the floor. Moreover, gravity acts upon everything: you cannot turn it off or shield it as you can other forces. For, whereas electricity and magnetism come in positive and negative varieties which cancel out, the gravitational 'charge' is mass and that only comes in positive doses. And it is this that allows gravity to rule in the domain of the very large. For when astronomically large bodies of matter accumulate, the net effects of the other charges in Nature tend to cancel out because they exist in positive and negative varieties. Mass, by contrast, just accumulates in the positive sense and eventually wins out despite its intrinsic weakness. Because of the weakness of gravity over the dimensions of physics laboratories, it is very hard to determine the form of the law of gravitational attraction very accurately there, whereas over the dimensions of the solar system the effects of uncertainties would be far more overt. If you look at the back pages of a physics textbook where they list the values of the constants of Nature for use in calculations, you will find that Newton's gravitational constant is specified to far fewer decimal places than any of the others. We could easily have failed to see some new force of Nature that had its most overt effects over these intermediate distances, greater than the atomic scale but smaller than the planetary. Of course, it is rather mysterious, although not impossible, that there should arise such unusual effects upon these, and only these, dimensions, which coincidentally lie so close to the scales of human experience.

The scenario of the 'fifth' force reminds us that there could well exist additional fundamental forces of Nature whose effects we have yet to recognize or perhaps even witness. Should we regard it as suspicious or merely fortuitous that all the forces of Nature are big enough for us to possess the expertise to detect them after a few thousand years of study? Does it not seem more probable that there exist additional forces of Nature that are intrinsically very weak, or highly selective in the things that they act upon, or which have a minute range? Such forces may well exist. They need not play any great role in the structure of the everyday world, or even the world of the present-day high-energy physicist, but their presence totally determines the form of the ultimate Theory of Everything that we seek. The number and nature of these ghostly forces determine the size and form of the ultimate symmetries of Nature. In order fully to unite them with the known forces, some constraint will have to exist upon the behaviour of the known forces that we may not as yet suspect. A scenario in which such a problem arises is equivalent to the over-arching symmetry problem, that is, whenever one has found some 'ultimate' symmetry that accounts for all the known interactions and particles of Nature, it is always possible to embed this in an even larger, grander, pattern populated by additional particles governed by new forces of Nature. This is a type of infinite regress tantamount to finding more and more elementary particles of matter at every level one probes. We have to hope that the possible laws that logical completeness allows are very small in number and possess symmetries which uniquely and completely specify the varieties and number of the particles or the stringy modes of vibration that can logically exist without destroying the structure of physical reality.

A mild version of this ghost-force problem arises in some of the string theories we have introduced earlier. Of the two special symmetries that these theories pick out for the world, one looks like the product of two identical patterns. As the Universe cools, the known forces of Nature can arise naturally from one of the copies of the pattern. But what happens to the other copy? There appears to be no reason why it should necessarily split up into a collection of different forces as well, although it could. Instead, it seems most natural that it remains in force as a sort of shadow world, where shadow images of all the known particles of matter interact very weakly, as though feeling only a feeble edition of the force of gravity. Such shadow matter could be threading its way around us all the time. The limits upon its presence and influence are rather weak and they display the vulnerability of our tidy worldview to influences that are not within the relatively small domain of strength and range that we can detect either directly or indirectly.

CHAPTER 5

Constants of Nature

But if thou wilt constant be
And faithful of thy word,
I'll make thee glorious by my pen,
And famous by my sword.
— MARQUIS OF MONTROSE

THE IMPORTANCE OF BEING CONSTANT

I often wonder, when reading descriptions of the scientific process by sociologists, if this is how an atom would feel if it could read a quantum mechanics textbook.
— JAMES TREFIL

There is something attractive about permanence. We feel instinctively that things that have remained unchanged for centuries must possess some attribute that is intrinsically good. They have stood the test of time. Our religious beliefs have traditionally focused upon the confidence that can be placed in an unchanging Supreme Being whose invariance 'yesterday, today, and forever' is thus a guarantee for the future. And, despite the constant flux of changing events, we feel that the world possesses some invariant bedrock whose general aspect remains the same. Physicists like to believe this also. The equations that they use to encapsulate the laws of Nature contain certain invariant numbers that have become known as the 'constant of Nature'. To endow a quantity with the epithet 'constant of Nature' gives it an especially exalted status within the scheme of things.

One of the wonderful expediencies of the equations that underpin the scientific investigation of Nature is that they can be used to predict the future without having any understanding of why these constant numbers possess the particular values that they do. We can simply measure them. If their values are slightly revised by better measurements, then no gross change occurs to the general form of the solutions to the equations. They are also disjoint from the nature of the initial conditions. Of course, we must be a little wary of this fortunate state of affairs. If constants of Nature arise as proportionality parameters in a particular, albeit useful, way of representing the world on paper, then these constants may be merely artefacts of the type of representation chosen. Perhaps there exist alternative ways of representing the physical world which lead to different invariant quantities? Certainly, the history of science has seen steady progress in making what once was arbitrary and complicated in our description of things seem increasingly compelling and simple. More often than not, this simplification occurs because quantities previously regarded as separate constants of Nature are found to be related or are discovered to be composed of combinations of other more basic constants of Nature. Each really major advance in physical science goes hand in hand with a revision or extension of our understanding of some constant of Nature. Newton's discovery of a universal law of gravity over three hundred years ago saw the introduction of a constant that now bears his name and specifies the intrinsic strength of the force of gravity in the Universe. The great novelty of this quantity for Newton and his followers was the fact that the measure of the intrinsic strength of gravity should indeed be a constant everywhere and everywhen. It linked together such superficially diverse phenomena as falling apples and moving planets.

Newton's constant of gravitation was the first of the modern constants of Nature to be identified. Its discovery had ramifications for other branches of philosophy and theology. Some of Newton's contemporaries pointed to its very universality as evidence for a single Authorship of the physical universe. It was often those of a unitarian religious persuasion, like Newton himself, who stressed this connection most forcefully.

As science has progressed and become more searching in the questions that it addresses to the world of reality, it is no longer content to regard these constants of Nature as standards that we can know by measurement alone. Even those negativists who at the turn of the century thought the work of physics was done save for the increasingly accurate measurement of the constants of Nature did not envisage that there might exist the possibility of calculating their values. They did not have such a programme on their agenda. The modern seekers after the Theory of Everything believe that ultimately some deep principle of logical consistency will permit these constants of Nature to be determined by simple counting processes. For these seekers and their forerunners, who can be found at every stage of the development of twentieth-century physics, the ability to predict these pure numbers is the real touchstone of a Theory of Everything. A theory that could successfully predict or explain the value of any constant of Nature would attract the attention of every living physicist. The truth of this statement is readily appreciated by those scientists who receive a large amount of mail from misguided members of the public announcing the discovery of their new 'Theory of the Universe' (the author has received two during the last week alone). Such proposals have two common factors (aside from the curious fact that in my experience they originate without exception from men rather than women): they aim to show Einstein was wrong in some way and they are totally committed to the deduction of the numerical values of the constants of Nature from some sequence of mysterious combinatorical juggling that occasionally incorporates considerations as abstruse as the dimensions of the Great Pyramid or the interpretation of the Jewish cabbala. The first of these factors has an obvious psychological motivation. Einstein is perceived as the twentieth-century scientist *par excellence*, and hence it is fondly imagined that, by catching him out on some point, the new author would be hailed as the new scientific Messiah, greater than Einstein. But the second factor is the more revealing. The manner in which the work of eccentrics is focused upon the computation of constants of Nature is a measure of the extent to which this quest is regarded as an ultimate goal of modern physics. It is clear-cut with an easily advertised answer. But where did this popular perception of the explanation of constants of physics as a Holy Grail for the physicist come from? I believe the answer is to be found in some of the work of scientists in the first half of the twentieth century that was extensively popularized at that time.

FUNDAMENTALISM

The great Arthur Eddington gave a lecture about his alleged derivation of the fine structure constant from fundamental theory. Goudsmit and Kramers were both in the audience. Goudsmit understood little but recognised it as far fetched nonsense. After the discussion, Goudsmit went to his friend and mentor Kramers and asked him, 'Do all physicists go off on crazy tangents when they grow old? I am afraid.' Kramers answered, 'No Sam, you don't have to be scared. A genius like Eddington may perhaps go nuts but a fellow like you just gets dumber and dumber.'
— M. DRESDEN

To appreciate fully the emphasis that was to come in the first half of the twentieth century, it should be borne in mind that at the close of the nineteenth century the centre of gravity of physics lay in Germany. But German physicists had inherited the philosophical legacy of Kant and this coloured many of their expectations concerning the ultimate capability of human scientific investigation. The known laws of physics were seen as the creations of the human mind and were to be distinguished from the true nature of things. Most textbooks carried an introductory discussion of the philosophy of science which stressed the perspective of Kantian idealism. In this underlying climate, there emerged a variety of views regarding the goal of explaining the entirety of the physical world and the constants of Nature that defined its gross form. On the one hand, there were those like Einstein who believed that the process of describing Nature by the laws of physics was a convergent one. There would always be elements of the current description that were inadequate, pieces of the true story that had been omitted. This ongoing process of revision which we call 'scientific discovery' might none the less have no end, for Einstein conceived of the general theory of relativity as only one further iteration towards the ultimate truth which lies at an unattainable asymptotia, for

> however we select from nature a complex [of phenomena] using the criterion of simplicity, in no case will its theoretical treatment turn out to be forever appropriate... But I do not doubt that the day will come when that description [the general theory of relativity], too, will have to yield to another one, for reasons which at present we do not yet surmise. I believe that this process of deepening the theory has no limits.

Soon after these words were written, Einstein began work on his infamous 'unified field theory', which was his vision of a Theory of Everything uniting his theory of gravitation with the laws of electromagnetism. If these 'two realities which are completely separated from each other conceptually' could be joined as 'one unified conformation', then 'the whole of physics would become a complete system of thought'. Later, all his intellectual energies were focused upon a search for 'theories whose object is the *totality* of all physical appearances'. Besides achieving a deeper description of the world in a unified way, Einstein also believed that a theory of this exalted type would resolve the uncertainties of the quantum theory that exercised him so greatly and also resolve the incongruity of the prediction that there be a beginning to the Universe, which his general theory of relativity indicated. This belief in the unity of Nature meant that Einstein's attitude towards the constants of Nature was that there should remain none whose values were not precisely explained by the internal consistency of any unified Theory of Everything, for

I cannot imagine a unified and reasonable theory which explicitly contains a number which the whim of the Creator might just as well have chosen differently, whereby a qualitatively different lawfulness of the world would have resulted....A theory which in its fundamental equations explicitly contains a constant [of Nature] would have to be somehow constructed from bits and pieces which are logically independent of each other; but I am confident that this world is not such that so ugly a construction is needed for its theoretical comprehension.

Einstein perceives the values of any unspecified constants of Nature as divine inputs that are required over and above those of the laws of Nature and the starting conditions for the Universe in order to specify the Universe uniquely. It is not easy to imagine how this could be true. Whilst one can envisage boiling down all the constants of Nature to an irreducible set of one or two pure numbers that characterize something about the size of the Universe and something like the string tension which tells us about a symmetry governing all the forces of Nature, as yet there is no inkling as to how this number could be reduced to zero. To achieve this would require the constants of Nature to be uniquely and completely prescribed by the form of the laws of Nature themselves.

Not all of Einstein's early contemporaries shared his vision of a constantless ultimate picture of Nature. Some, like Max Planck, saw physical science as an essentially inductive enterprise that could never give way to some ultimate Theory of Everything that was arrived at by pure deduction. Hence, for him, there could be no attainable all-encompassing theory that explained the values of all the constants of Nature. Planck was far from being a Kantian idealist and actually saw the hallmark of progress in science as being the systematic progress 'towards as far-reaching a separation as possible of the phenomena in the external world from those in human consciousness'. He characterized the search for a Theory of Everything as the quest for a 'single world formula'. Others, like the instrumentalists Pierre Duhem and Percy Bridgman, regarded the promised Planckian separation of scientific description from human conventions as unattainable in principle, since they viewed the constants of Nature as entirely the result of a particular framework of human explanation being imposed upon an unknowable reality.

Sir Arthur Eddington was the greatest astrophysicist of the pre-World War II era. He had founded the systematic study of the structure of stars, significantly furthered our understanding of the motions of stars in the Milky Way, and provided his contemporaries with the finest exposition of Einstein's new theory of gravitation, while playing a key role in its experimental confirmation. Eddington was also a Quaker of deep conviction and it is interesting to wonder how the Quaker notion of the 'inner light' may have played some role in his scientific thinking. Despite being a shy man who spoke haltingly and reluctantly in public, Eddington had a golden pen. His writings on science for the scientist and for the wider public are amongst the finest ever penned and are still widely read today. His ability to write with unparalleled charm and lucidity meant that his popular books were more widely read than any other expositions of contemporary science. They exerted an enormous influence upon philosophers and others who were first introduced to developments in physical science through the medium of his writing. And what Eddington's readers found interwoven with his eloquent exposition of the facts was an underlying philosophy of science that differed dramatically from that offered by any other leading scientist of the time (and more still from any English scientist of any time). Eddington was close to being a Kantian. He regarded the part played by the human mind in constructing our picture of the physical world as being fundamentally irreducible. But, whereas some idealists would have used this as an excuse to look no further into the nature of things, Eddington saw the origin of laws of Nature in the human mind as a guarantee of their ultimate rationality. He asks whether it is not 'possible that laws which have not their origin in the mind may be irrational' to the extent that 'we can never succeed in formulating them'. Motivated by Russell and Whitehead's monumental reduction of mathematics to the most basic propositions of logic, Eddington sought to reduce physics to its lowest terms and so discover whether the success of our schemes for explaining the workings of the world owes everything to some intrinsic simplicity of Nature or to the fact that they are creations of our minds. Throughout the latter part of his life, until his death in 1944, Eddington worked upon what he called his *Fundamental Theory*. Only pieces of it were published in his lifetime. It was an attempt to arrive at a Theory of Everything that had as its primary goal the explanation of the numerical values of all the constants of Nature in terms of elaborate counting arguments. The underlying philosophy of this work has been described by Whittaker (one of his scientific biographers) as the belief that

all the quantitative propositions of physics, that is, the exact values of the pure numbers that are constants of science, may be deduced by logical reasoning from qualitative assertions without making any use of quantitative data derived from observation.

This, you will recognize, is the complete opposite of Planck's view of the matter. Eddington believes that the partial human contribution, that Kant would have regarded as total, allows us to contemplate a completely self-consistent description of the world without leaving any quantities to be determined by observation alone.

Eddington's attempts to explain the values of the fundamental constants of Nature failed totally in the judgement of other physicists. They regarded this work as divorced from real physics, amounting to little more than wishful juggling of numbers to get the answers he wanted. It is fair to say that with the benefit of the forty years during which it has been possible to examine Eddington's posthumous manuscript on the subject that nothing of any value to science emerged from it save for the clarion call to explain the values of the constants of Nature. It was Eddington's work, and more particularly his reporting of its curious interim results in his marvellous books, that is responsible for so many amateur scientists who try to follow in his slightly misguided footsteps and explain the values of constants of Nature by arithmetical gymnastics. But what is most curious to the modern physicist reading Eddington's attempts to derive the constants of Nature is that the logic employed has no point of contact with any other area of science. Indeed, even during his own day, Eddington had great difficulty impressing other scientists with the seriousness of this work. Many spoof articles were written parodying his approach and he was more often than not accused of being obscure and unintelligible. He found this exasperating, especially when the work of Cambridge colleagues like Dirac was regarded as of such great importance. Of his own attempts to explain the constants of Nature, he wrote to a friend:

> I am continually trying to find out why people find the procedure obscure ... I cannot seriously believe that I ever attain the obscurity that Dirac does.

Eddington's was a research programme that failed. Like Einstein's, it was premature. We simply did not know enough about those ingredients that must be included in a Theory of Everything to embark upon its construction. Yet more than any other it brought into the limelight the challenge of explaining the constants of Nature. Rather as the Design Arguments of the nineteenth century teed up the facts for Darwin to explain by the process of natural selection, so Eddington set up the problem of the values of the constants of Nature as a target for the sharp-shooters of the future.

WHAT DO CONSTANTS TELL US?

A marvellous newtrality have these things mathematicall and also a strange partici-
pation between things supernaturall, immortall, intellectuall, simple and indivisible,
and things naturall, mortall, sensible, componded and divisible.
—JOHN DEE

We have been considering the importance that physicists have traditionally
attached to the values of the constants of Nature, but what role do these
constants play in the Universe? Why are they considered so important? Some
perspective can be gained by considering first the world of atoms and molecules. These entities are not elementary particles, rather they are composites of
many particles held in balance by opposing forces. The sizes of these structures
determine the density of matter and the arrangements of electrons in atoms
generate the entire range of chemical properties of matter. Yet despite the vast
complexity of everything made out of atoms and molecules, together with the
vast range of properties that straddle the states of matter from gases to liquids
to solids, the gross features of this entire world of materials is determined by
the values of just two numbers. The numbers in question are: the mass of the
proton (which is the nucleus of the hydrogen atom) divided by the mass of the
electron,

$$\frac{\text{proton mass}}{\text{electron mass}} = 1836.152 \ldots,$$

and a quantity that has become known as the 'fine-structure constant'. This is
the square of the electric charge on a single electron divided by the product
of the speed of light and Planck's constant of quantum theory. This particular
abstruse combination is taken because it produces a pure number. Its peculiar
value of $1/137.036 \ldots$ is obtained by combining the measured values of the three
constants that comprise it. We do not know why these two numbers take
the precise values that they do. Were they different, our Universe would be
different, perhaps unimaginably different.

If we look beyond the Earth at the structure of the solar system, then
the chemical forces are joined by the force of gravity in determining the
gross features of things. The strength of gravity is determined by Newton's
gravitational constant, and from this quantity we can determine another pure
number like the fine-structure constant but with the square of the electron
charge now replaced by the product of Newton's constant and the square of
the mass of the proton. This number, the gravitational-structure constant, has
a tiny value equal to $5.9041183 \ldots \times 10^{-39}$. Its smallness compared with $1/137$ is

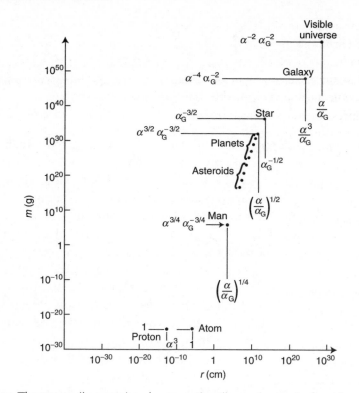

Figure 5.1 The masses (in grams) and average sizes (in centimetres) of a wide selection of all the principal objects we know to exist in the Universe. The composite structures are equilibrium states between different forces of Nature and their approximate sizes are determined by the fine-structure constant $\alpha = 1/137$ and the gravitational-structure constant $\alpha_G = 5.9 \times 10^{-39}$, which we introduced in the text; the dependence of the mass and size upon these two quantities is indicated for each object.

telling us that the chemical forces of electromagnetic origin are far stronger than those of gravity. Indeed, gravity is utterly irrelevant to the structure of atoms. It is there, but its effects are so minute compared with the electric forces between the protons and electrons that they can be totally ignored in all practical considerations of chemistry and nuclear physics. The sizes of all the astronomical bodies from the scale of asteroids up to stars are determined by the relative values of the fine- and gravitational-structure constants alone. This is shown in Figure 5.1, from which one can see the effects of a change in the values of the fine- and gravitational-structure constants. The sizes of planets and stars are not random accidents or the pre-programmed result of particular initial conditions at the Big Bang. Rather, they arise as equilibrium states between opposing forces of Nature. These forces come into balance only when the aggregate of particles involved attains a certain size. In cold bodies, like the Earth, the compressive force of gravity, trying to compress everything to a smaller size of higher density, is opposed by a quantum mechanical effect known as the exclusion principle. Particles like protons or electrons occupy microscopic niches into which only one particle is allowed to sit. Any attempt to compress matter so that more than one particle would be squeezed into each niche is met by a resisting force. The balance between this force and the inward push of gravity results in the large, stable, cold bodies we see in the solar system.

Stars are different. A star is a body that is massive enough for the gravitational compression at its centre to produce a temperature great enough for nuclear reactions to occur spontaneously. When this ignition temperature is achieved, nuclear reactions in the central region will produce an energy outflow that is eventually radiated away from its surface in the form of heat and light. The star is kept in equilibrium by the balance between its internal pressure and gravity. This is a stable balance because if the gravitational force were made slightly greater then the star's centre would be squeezed a little more, so producing faster nuclear reactions and hence a corresponding extra pressure pushing outwards. Thus a balance is quickly restored.

The fact that so many of Nature's most important creations owe their gross size and structure to the mysterious values of the constants of Nature places our own existence in a new and illuminating perspective. We can see how the conditions necessary for our own existence are contingent upon the values taken by the constants. At first one might imagine that a change in the value of a constant would simply shift the size of everything a little, but that there would still exist stars and atoms. However, this turns out to be too naïve a view. It transpires that there exist a number of very unusual coincidences regarding the values of particular combinations of the constants of Nature which are necessary conditions for our own existence. Were the fine-structure constant to differ by roughly one per cent from its actual value, then the structure of stars would be dramatically different. Indeed, there is every reason to suspect that we would not be here to discuss the matter. For the biological elements like carbon, nitrogen, oxygen, and phosphorus are produced during the final explosive death throes of the stars. They are blown out into space where they become incorporated into planets and, ultimately, into people. But, carbon, the crucial biological element which we believe to be essential for the spontaneous evolution of life, should really only exist as the minutest trace element in the Universe instead of in the healthy abundance that we find. This is because the explosive nuclear reactions that make carbon in the late stages of stellar evolution are typically rather slow at producing it. However, there exists a remarkable coincidence of Nature that allows carbon to be produced in unexpected abundance.

Carbon originates in the Universe via a two-step process from nuclei of helium, or alpha particles as we usually call them. Two alpha particles combine under stellar conditions to make a nucleus of the element beryllium. The addition of a further alpha particle is necessary to transform this into a carbon nucleus. One would have expected this two-step process to be extremely improbable, but remarkably the last step happens to possess a rare property called 'resonance' which enables it to proceed at a rate far in excess of our naïve expectation. In effect, the energies of the participating particles plus the ambient heat energy in the star add to a value that lies just above a natural energy level of the carbon nucleus and so the product of the nuclear reaction finds a natural state to drop into. It amounts to something akin to the astronomical equivalent of a hole-in-one. But this is not all. While it is doubly striking enough for there to exist not only a carbon resonance level but one positioned just above the incoming energy total within the interior of the star, it is well-nigh miraculous to discover that there exists a further resonance level in the oxygen nucleus that would be made in the next step of the nuclear reaction chain when a carbon nucleus interacts with a further alpha particle. But this resonance level lies just *above* the total energy of the alpha particle, the carbon nucleus, and the ambient environment of the star. Hence, the precious carbon fails to be totally destroyed by a further resonant nuclear reaction. This multiple coincidence of the resonance levels is a necessary condition for our existence. The carbon atoms in our bodies which are responsible for the marvellous flexibility of the DNA molecules at the heart of our complexity have all originated in the stars as a result of these coincidences. The positioning of the resonance levels are determined in a complicated way by the precise numerical values of the constants of physics.

There are many other examples of this ilk. At almost every turn, the conditions necessary for the evolution of any form of complexity in the Universe exploit the occurrence of crucial coincidences between the values of the constants of Nature. Some have taken this to be of great theological import, regarding it as a form of divine fine-tuning of the Universe so as to make the evolution of life a certainty. Such arguments are reminiscent of the Design Arguments of the natural theologians of the past. Yet we cannot claim anything more than that such life-supporting coincidences are necessary for the evolution of life as we know it.

The evolution of life and mind is beset by evolutionary dead-ends at every stage. There are just so many ways in which life can fail to evolve in a complex and hostile environment that it would be sheer hubris to suppose that, simply given enough carbon and enough time, anything is possible. Moreover, to suppose that life must result from the requisite mixture of chemicals is just the sort of teleological attitude that the biologists so rightly decry. There is no reason why life has to evolve in the Universe. Such complex step-by-step processes are not predictable because of their very sensitive dependence upon the starting conditions and upon subtle interactions between the evolving state and the ambient environment. All we can assert with confidence is a negative: if the constants of Nature were not within one per cent or so of their observed values, then the basic building blocks of life would not exist in sufficient profusion in the Universe. Moreover, changes like this would affect the very stability of the elements and prevent the existence of the required elements rather than merely suppress their abundance.*

It is not easy to interpret this state of affairs. Let us grant that there do exist vital coincidences between the values of constants of Nature whose existence is necessary for our own. Let us then go further and suppose, for argument's sake, that these coincidences are necessary for the evolution of any form of self-aware complexity of the sort that we call 'conscious life'. What, one may ask, would be the impact of a Theory of Everything that successfully explained the values of all the constants of Nature? If that theory was unique, permitting only one possible deal of the constants, then all we can say is that we were rather fortunate. Any deeper conclusions of a metaphysical nature must necessarily be speculative and very likely impossible to exclude. But, if the Theory of Everything shows the values of the constants (or at least some of them) to possess some random element that depends upon the particular events that unfold locally, or that the quantities that we take to be constants, can all in principle (and maybe also in practice) vary randomly in space, then in an infinite universe there would necessarily arise regions that would possess combinations of the constants suitable for the subsequent evolution of complexity. We must, of course, inhabit one of those cosmic oases of life.

Considerations such as these show us why scientists would love to be able to explain the values of the constants of Nature. They explain why just about everything that does exist can exist. Yet we can also see again why the concept of a constant of Nature has proved such a fruitful one. Our inability to explain

* Readers interested in discovering all the details of these coincidences and their consequences should consult *The Anthropic Cosmological Principle* by the author and F. J. Tipler.

why the fine-structure constant has a value close to $\frac{1}{137}$ rather than $\frac{1}{145}$, say, does not inhibit us from using the concept of the fine-structure constant and arriving at an understanding of how its value determines other things. This is a manifestation of a general property of the world which helps make it relatively intelligible to us. There exists a form of hierarchical structure in Nature which permits us to understand the way in which aggregates of matter behave without the need to know the ultimate microstructure of matter down to the tiniest dimensions. If the deep logic of what determines the value of the fine-structure constant also played a significant role in our understanding of all the physical processes in which the fine-structure constant enters, then we would be stymied. Fortunately, we do not need to know everything before we can know something.

But we do not have to appeal to the esoteric worlds of elementary particles or astronomy to appreciate the importance of constants of Nature. It is vital to the functioning of modern technology and communications that there exist precise standards of time measurement. Accordingly, most developed countries have special 'national laboratories' part of whose job is to maintain rigorous measures of time, length, and mass, along with other measurement standards. In the United Kingdom, this duty falls to the National Physical Laboratory, whilst in the USA it is met by the National Bureau of Standards in Washington. What these institutions require for timekeeping purposes is a standard of time that is absolutely fixed. Such a standard can then be used as the basis from which to calibrate all other secondary measures of time. Suppose, like the ancients, one thought that a sand clock (an egg-timer, for example) was up to the task. This device uses the force of gravity to tell the time. It exploits the fact that everything falls with the same acceleration under the force of the Earth's local gravitational field. But clearly such a device is an impossible terrestrial standard, let alone a universal one. The size of the hole through which the sand escapes will always be slightly different from device to device; the granularity of the sand will differ from sample to sample; the texture of the surface down which the sand slides and the angle of its inclination will differ: all of these factors make each sand clock different from its rivals. There is no unique relationship between the change in the fall of sand and the passage of time. One could try to overcome this by employing a pendulum clock. Again, this device relies upon the local force of gravity to dictate the period of each oscillation of the pendulum. But the period of each swing of the pendulum also depends upon the length of the pendulum. Each pendulum will therefore be slightly different. Moreover, the effective pull of gravity at the Earth's surface varies as one goes from the equator to the poles because of the rotation of the Earth on its axis and the slight oblateness in the shape of the Earth that this creates. If one took the clock to another planet, then the strength of gravity at its surface would differ from that on the Earth and the clock would tick at a completely different rate when compared with an exact counterpart on the Earth. The bigger the planet, so the faster would a pendulum of the same length swing under its gravity (increasing roughly as the square root of the radius of the planet in fact). If we were to get a little more sophisticated, then we could turn to an electric clock of the sort that is found in most households. This is far more accurate than a pendulum clock and it achieves its accurate periodicity by reference to the alternation cycle of the domestic a.c. electricity supply. Nevertheless, although this supply maintains a fairly steady frequency close to fifty cycles per second, it is subject to unpredictable variations that differ from place to place and from time to time. Such a standard could never be truly universal, although it is perfectly adequate for most everyday purposes. A true standard requires us to find some way of defining what we mean by one unit of time which is the same for everyone no matter where or when they are observing. This desire for universality naturally moves us to seek some time standard that is determined by the constants of Nature alone. And this is indeed how modern absolute time standards are defined. They avoid the use of any characteristic of the Earth or its gravitational field and focus instead upon the natural oscillation frequencies of certain atomic transitions between states of different energy. The time for one of these transitions to occur in an atom of caesium is determined by the velocity of light in a vacuum, the masses of the electron and proton, Planck's constant, and the charge on a single electron. All these quantities are taken to be constants of Nature. A time interval of one second is then defined to be a certain number of these oscillations. Despite the esoteric nature of this definition of time, it is a powerful one. It should allow us to communicate precisely what length of time we were talking about to the inhabitants of a distant galaxy. Whilst Andromedans would presumably have no knowledge of what a year or a day are, because these are units of time that are properties of the motions in our particular solar system (the day is the time for one complete revolution of the Earth and the year the time for the Earth to complete an orbit of the Sun, and neither is truly constant in fact), they would, if they were talking to us using radio signals, have to be familiar with the concepts involved in the definition of the constants of Nature. The inevitability of their knowledge of these quantities, we might argue, ensures that we will have much in common. Or does it?

VARYING CONSTANTS

There is nothing in this world constant but inconstancy.
—JONATHAN SWIFT

There are two 'ifs' here. Would extraterrestrial civilizations have discovered the same constants of Nature? And are these so-called 'constants' really constant anyway? The first question is both a philosophical and a sociological one. If one adopts a realist view of science, then one holds that there is a true and unique structure to the physical universe which scientists discover rather than invent. The constants of Nature exist in a mind-independent sense. They are not concepts that the human mind has created simply in order to make sense of the facts. On this view, one could argue that any scientific or technological civilization would have to discover the same reality and the same basic concepts. They might use different symbols or use constants that were defined slightly differently for the sake of convenience, but they would none the less recognize our constants as fundamental and could readily translate their own constants of Nature into ours. This rather optimistic view actually lies at the heart of several proposals to search for extraterrestrial life. The messages to extraterrestrials borne by our spacecraft as they head out of the solar system having completed their mission as well as the frequencies on which radio signals are beamed out into space in case anyone is listening, both focus upon fundamental standards dictated by the constants of physics. It is implicitly assumed that these will be recognized by any technological society. But maybe this line of thinking is wrong. If our mathematics and physics are largely invented in order to describe some vastly deeper true reality, then we would not expect extraterrestrials to have taken the same path at all. Our scientific concepts could have developed in response to the social and practical problems that needed to be solved on planet Earth. The seemingly fundamental mathematical notions that form the bedrock upon which our science is founded may owe much to the primary concepts that our minds seem to accommodate most easily. These mental attributes are at least partially the result of an evolutionary process that is driven by the environmental characteristics of the Earth. On other worlds, this development would have been different. The results would have been different. We could have confidence in there being one basic common factor only. There should be a close correspondence between the image of reality that a successful creature has and the true nature of those aspects of reality necessary for survival. A serious mismatch between image and reality in this department will inhibit the likelihood of evolutionary success. Yet this still leaves enormous scope for divergence. For example, on our planet, there is a fairly transparent atmosphere that allows us to see large numbers of stars at night. If a planet were dark and obscured by dense cloud, then sound would have been a more effective means of primitive communication than light. Such a difference might well bias the future development of any advanced civilization to stress the study of sonic rather than electromagnetic phenomena. An all-embracing aspect of this whole question is the status of the mathematical language in which all our deductions about constants of Nature are framed. Is this an invented or a discovered facet of the Universe? This is such a large topic that we shall be returning to discuss it more fully in the final chapter.

What of our second question? Are the constants really constant? So far, we have assumed, as most physicists do, that quantities like Newton's gravitational constant, the charge of an electron, or the fine-structure constant are truly constant. This is not merely a pious hope. One can test this assumption in various ways. When one observes distant astronomical objects, like quasars, we are seeing them as they were billions of years ago because of the enormous period of time that must pass before the light signals they emit reach our telescopes on Earth. This time-lag allows us to test whether the constants of physics which dictate the relative properties of different types of light emitted by the distant source are identical to their counterparts on Earth today. If certain constants of physics had differed in value in the distant past, then the differences would have shown up unless they are smaller than one part in a hundred billion. We also know that if quantities like the fine-structure constant or the gravitational-structure constant varied in the past then the behaviour of events in the early stages of the Universe would have been very different. In particular, the beautiful agreement between the observed abundances of hydrogen, helium, deuterium, and lithium in the Universe and what is predicted to emerge from the Big Bang when the Universe was just a few minutes old would be destroyed. The consequences of variations in the constants of Nature that play a role in these processes are typically constrained to be smaller than between one part in ten billion and one part in one thousand billion in order that the observed abundances are not significantly changed.

There are circumstances in which we might expect those quantities that we call constants of Nature to exhibit a variation in time or space. We observe there to exist three dimensions of space, but particle physicists have discovered that the most elegant and complete theories of elementary-particle processes, in particular the string theories of which we spoke in the last chapter, predict that there are many more than three dimensions of space (perhaps a further six, or even another twenty-two in some cases). To square such a state of affairs with what we see, it is required that all but three of the dimensions of space be microscopically small. Let us suppose that this is indeed the case. Then the true constants of physics are those that determine the nature of the whole of space, not just our three-dimensional slice of it. Moreover, this has the consequence that if we examine those quantities which, in our three-dimensional subset of the world, we have come to call the constants of Nature then we will find them to change at the same rate as the average size of any additional dimensions of space. The observations of unchanging constants tell us that if there are any extra dimensions of space then today they are inert to fantastically high precision.

This scenario of extra dimensions of space is more than merely a wild speculation because the string theories of which we spoke in the last chapters do indeed possess many additional dimensions. In fact, the miraculous mathematical properties which enable them to cure the infinities of the point-particle theories appear to require that there exist either nine or twenty-five dimensions of space. In a cosmological setting, they require us to envisage the earliest stages of the Universe, when the stringiness of things is dominant, as one in which all these dimensions exist on an equal footing. Then, for some unknown reason, there must have come a parting of the ways. Three of the dimensions of space must have expanded to become the present-day visible universe, fifteen billion light years in extent. The remainder must have been held static on a microscopic scale. How this was achieved and why three and only three dimensions have escaped from this perpetual imprisonment remains a mystery.

An interesting point of principle emerges here. If there are additional dimensions of space, then the true constants of Nature are defined over the totality of the dimensions of space. Those that we see in three dimensions may not therefore be truly fundamental. They may not be the constants that the ultimate Theory of Everything would tell us. If so, we would have to unravel the entire process by which three of the dimensions are not only large but getting larger as our observations of the cosmic expansion indicate, whilst the others stay small and static. This process may not be determined by the laws of Nature. It may possess intrinsically random elements at the quantum gravitational level, which could even make it vary from place to place in the Universe. Were this the case, then the values of the constants that remain in the three large dimensions of space receive values that are at least partially random in origin.

For many years, the quest to find evidence for changes in the traditional constants of Nature drew a blank. All that was obtained were stronger and stronger restrictions on the magnitude of any changes from direct laboratory experiments and astronomical observations. Things began to change in 1999 when a group of us started to exploit the extraordinary power of new telescopes like the Keck on Hawaii to study the light from distant quasars in a new way. John Webb, Michael Murphy, Victor Flambaum, Vladimir Dzuba, Chris Churchill, Jason Prochaska, Art Wolfe, and I have applied a new technique to analyse light from distant quasars. We look at the separation between lines caused by the absorption of quasar light by different chemical elements in clouds of dust in between the quasar and us. These separations depend sensitively on the value of a particular constant of Nature, the 'fine-structure constant', at the redshift (or the distance away from us) where the absorption occurs. By comparing the observed values with their separations here and now in the laboratory we have a probe which can tell us whether this 'constant' can have changed over twelve billion years. This method has now been applied to observations of 147 quasars. The results gathered and analysed over two years have proved to be unexpected, and potentially far-reaching. We find a persistent and significant difference in the separation of spectral lines in the distant past compared with the separation of the same lines when measured in the laboratory today. The complicated 'fingerprint' of shifts matches that expected if the value of the fine-structure constant was *smaller* by about six parts in a million at the time when the quasar light was first emitted. Many other sets of observations are now being scrutinized by astronomers to see if this tantalizing change in one of Nature's fundamental 'constants' is real, or the result of some unappreciated complication in the data-gathering process. During the past year, other astronomers have reported evidence consistent with similar tiny shifts in the ratio of the electron and proton masses.

Simple theories can be created which include the variation of the fine-structure constant. Remarkably, any variation is fairly strongly constrained by the need for it to conserve energy and momentum and by the effects of the force of gravity on the changes. We find that the 'constant' can only vary very slowly—increasing as the logarithm of the age of the universe—during a particular era of the Universe's history, from when it is about 300,000 years old until it is about 9 billion years old (today it is 13.7 billion years since its expansion appears to have begun). At other epochs, the fine-structure 'constant' remains constant. This ensures that the observed variation in quasars, if real, actually has no other observable effects on events in the very early history of the Universe or in laboratory experiments, which are not as sensitive as the quasar observations. The only place where effects might show up is in the past geochemical history of the Earth. Two billion years ago, in Gabon in West Africa, a complicated sequence of geological accidents created a natural nuclear chain reaction beneath the Earth's surface at a site that is now a uranium mine. The nuclear reaction sequence that is now fossilized in the radioactive elements there in the mine site is a very unusual one that relies on a very unusual coincidental location of an energy level in the samarium nucleus that enables it to capture neutrons far more effectively than would have otherwise been the case. The energy level must have been present with the same special value to very high precision two billion years ago. This severely constrains any variation of the fine-structure constant over the past two billion years, but unfortunately this lies in the period of time when we do not expect the fine-structure 'constant' to be varying because the Universe is expanding too quickly, and dominated by its dark energy content.

THE COSMOLOGICAL CONSTANT

Nothing. Nothing is an awe-inspiring yet essentially undigested concept, highly esteemed by writers of a mystical or existentialist tendency, but by most others regarded with anxiety, nausea, or panic.
—THE ENCYCLOPAEDIA OF PHILOSOPHY

As we might expect, the modern candidates for a Theory of Everything seek to say something about the constants of Nature and their values. The first candidate, the string picture of elementary particles with its new notion of what are the most elementary entities in Nature and its discovery of the power of mathematical consistency in narrowing down the candidates for the grand symmetry that governs everything, maintains that the values of the constants of Nature are contained within the Theory. They must be buried deep within the mathematics in some intricate way. Unfortunately, there has as yet been found no way to extract this information from the theory. Indeed, as we have just seen, this extraction might be only the first stage in a difficult journey, because it is one thing to discover what the Theory has to say about constants living in nine or ten dimensions of space, but quite another to discover how this information then filters down to determine the values of measured constants in our three-dimensional world. Another problem that we have to bear in mind is the fact that what emerges as a fundamental constant from a Theory of

Everything may not be quite the same as one of the quantities that we are in the habit of calling a constant of Nature. We may be somewhat removed from the ultimate constants of Nature because of the way in which our understanding of the physical world has been obtained in the relatively low-energy environment which is a necessary prerequisite for the evolution and survival of living things. Yet modern physics points always towards higher and higher energies and environments that are maximally removed from our own intuition and experience for the answers to our deepest questions. In string theory, for example, the most fundamental quantity appears to be the string tension rather than one of the conventional constants of Nature. At first, one might think this situation to be but one of mild inconvenience. Suppose that we were able to unravel the process by which many dimensions divide into three that expand to become large and the rest that stay small and we can calculate the numerical values of the underlying constants of the Theory of Everything. How could we be stymied? To answer this, we have to recognize that there are constants of Nature and there are constants of Nature. Whilst we tend to lump them all into the same category for ease of discussion, there appear to be some which are more fundamental than others. Those that are most fundamental we would expect to be determined totally by some internal logic of the Theory of Everything, but there are others which appear to take on contributions from particular processes that go on in the Universe which alter their values in ways that are unpredictable. The Theory of Everything might leave us some way from predicting the values of the observed constants of Nature.

Despite this possible complication, one would feel elated if a candidate Theory of Everything could predict even one constant of Nature correctly. Until quite recently, it was hoped that string theory might eventually be able to do this, and then we need only compare the prediction with observation to test the theory. But, in recent years, a rival Theory of (almost) Everything has emerged which has called into question the extent to which there are any truly fundamental constants that remain forever the same, unaffected by events in space and time. It is to this rival that we now turn.

There is one sort of constant that is somewhat easier to explain than all the others. It is one whose value is zero. In 1916, when Einstein first applied his new theory of gravitation to the Universe as a whole, he was greatly influenced by the philosophical prejudices of the past, which held there to exist a static absolute space upon which all the observed local motions took place. The very idea of the expansion of the Universe as a whole would have been a most peculiar and unacceptable one. Einstein discovered this expansion to be an immediate consequence of his theory, and so investigated how the theory might be modified in order to suppress the presence of expanding (or for that matter contracting) universes. He realized that this could be done by simply including in the theory a new constant of Nature that was permitted but not demanded by its mathematical development. This new constant, called the *cosmological constant*, acted as a long-range force opposing the force of gravity. A balance between the two forces was possible resulting in a static unchanging universe, which was an impossibility when the cosmological constant was omitted from the theory. However, it was subsequently shown that, while there can indeed exist such a static universe in principle, it could never persist in practice, because the equilibrium between the pull of gravity and the push of the cosmological constant is an unstable one. Like balancing a pencil upon its point, the slightest tremor in one direction or the other causes a one-way change. If Einstein's static universe were not perfectly balanced (and in the real world it never could be, because of changing fluctuations and non-uniformities, like ourselves, that we know to exist in the Universe), then it would soon start either to contract or to expand. As this situation became clear, interest turned to the study of an expanding universe and was inspired further by Hubble's monumental discovery in 1929 that the Universe *is* expanding.

Einstein's static universe was a dead duck and its inventor later regretted the introduction of the cosmological constant to sustain a static universe, calling it 'the biggest blunder of my life', because he missed the opportunity to make the greatest scientific prediction of all time: the expansion of the Universe. But the cosmological constant refused to make a graceful exit. It was all very well to say that it was no longer required, but there was really no good reason why it should be excluded from Einstein's equations. Many regarded it solely as an accoutrement to Einstein's new theory of gravity, and so it could be deleted on the grounds that it had no counterpart in Newton's classical theory of gravity. For, if it were retained, then Newton's theory, to which Einstein's reduces when the strength of all gravity fields are very weak and all motions occur at speeds far less than that of light, is no longer recovered. Instead of Newton's famous inverse-square law, which says that the force of gravity between two masses whose centres are separated by a distance d is

$$\text{Force} \propto 1/d^2, \qquad (*)$$

it is found that the law of force has the form

$$\text{Force} \propto 1/d^2 + \Lambda d, \qquad (**)$$

where Λ is the cosmological constant.

It is an interesting curiosity that Newton could indeed have arrived at this force law three hundred years ago. One of the problems that exercised him greatly, and delayed the publication of his *magnum opus*, the *Principia*, for many years was to justify his habitual assumption that the gravitational pull exerted by a spherical mass is identical to that exerted by an equal 'point mass' of zero extent located at its centre. This is known as the 'spherical property'. Eventually Newton established that this assumption is true for the inverse-square law of force (∗) that constitutes his law of gravity, but not for other hypothetical force laws like inverse cube or fourth powers of distance.

But, if Newton had asked the question 'What is the general force law such that the spherical property holds?', as the French mathematician Simon de Laplace was eventually to do, then he would have found that the law (∗∗) is the answer. So, if Newtonian gravity had been formulated with the spherical property as its foundational principle, then the cosmological constant would have been permitted but not demanded, exactly as Einstein found the situation to be in the foundation of the general theory of relativity.

What is the status of the cosmological constant today? Up until about 1997, the situation was not very engaging. There was no observational evidence for the existence of a cosmological constant at the level of observational sensitivity then existing. This meant that its numerical value had to be less than about 10^{-55} per cm^2, so small that most physicists suspected that it was really equal to zero. Eventually, they argued, we would find some simple principle in the Theory of Everything that would tell us that it had to equal zero.

Things fell out very differently. At the end of the 1990s two large astronomical research groups using a new observational technique started to find evidence that the cosmological constant was not zero it all. By studying the flaring and fading of large numbers of supernovae close to the edge of the visible universe, they were able to monitor the expansion of the Universe out to distances where the second term in the force law (∗∗) would come into play if Λ was large enough. They found that at large enough distances the expansion of the universe slowly changes gear from a state of deceleration, governed by an attractive gravitational force into one of acceleration driven by universal repulsion. This is exactly the behaviour expected of a cosmological constant. It appears to have a value close to about 10^{-56} per cm^2, and accounts for 72 per cent of the gravitating material in the Universe.

This is a very strange situation. The cosmological constant has a tiny value, but if it were a little more than ten times bigger then we could not be here. It would have started to accelerate the expansion of the Universe earlier in its history, and it would not have been possible for stars and galaxies to form.

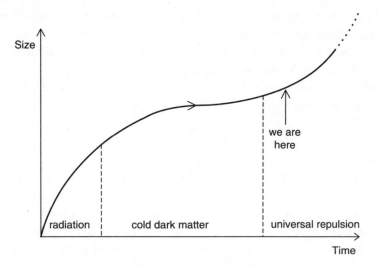

Figure 5.2 The change of the separations between distant points ('size') with time in an expanding universe in which the cosmological constant eventually starts to accelerate the expansion. This expansion trajectory is the best-supported picture of the history and future of our observable universe. The universe passes from an early period of expansion that is dominated by radiation into one where the expansion rate is driven by cold dark forms of matter; but then, when it is about three-quarters of its present-day size, it changes gear and begins to accelerate.

The expansion would have been too rapid for local gravitational clustering to beat. As it is, we are confronted with a picture of the Universe in which the expansion begins to accelerate after the Universe expands to about three-quarters of its present extent (see Figure 5.2).

The cosmological constant has been interpreted by particle physicists as a measure of the quantum vacuum energy of the Universe. In a quantum system the notion of a vacuum is a little different from our usual conception of such a state. It is not simply 'nothing at a'. Rather, it is what is left when everything that can be removed from the system has been removed: it is the state of lowest energy. This means that it need not correspond to a zero energy level and there can even be two (or more) different states with the same minimum energy. Moreover, the minimum energy state can change as time passes so that it ceases to be the minimum energy state after a while and the system quickly (or slowly) changes from one vacuum to another. Seen in this way, the small value of the cosmological constant that we observe today is telling us the vacuum energy of the Universe. Unfortunately, this does not really tell us why it takes the strange value that it does. All the theories of particle physics that have something to say about it predict that its value should be hugely bigger—as much as 10^{120} times bigger in most cases!

There is evidently a very big gap in our understanding of the vacuum energy of the Universe and the cosmological constant. So far, string theory has shed no light on it either. Indeed, it is possible that its value may not have an explanation in the conventional sense that physicists have come to expect. String theories turn out to allow a huge multiplicity of possible universes as outcomes of the Theory of Everything that they define. Many of the properties of the Universe that we are in the habit of regarding as fundamental can have a lesser status in these theories because they are not specified uniquely and completely by the Theory of Everything and the laws of Nature that it specifies. The number of vacuum states into which the Universe can fall seems to be stupendously large—10^{1500} is one estimate—and each of these states is characterized by a different set of values for many of the constants of Nature, a different value for the cosmological constant, and different structures for the astronomical universe. Worse still, they even allow different numbers of forces of Nature and, conceivably, different numbers of dimensions of space that become large. In a situation like this, all we can hope to determine from the Theory of Everything is something like the probability that a universe will turn out to possess a particular suite of forces, constants of Nature, and other properties. There will be no fundamental explanation for the observed value of a quantity like the cosmological constant at all. We can identify the range of values of a quantity like the cosmological constant for which it is possible for conscious 'observers' to exist. Obviously, we have to find that we are living in a universe that possesses constants and properties that fall within the life-supporting range *no matter how improbable they may be in the Theory of Everything a priori*. We might be able to determine if an observed feature was typical or very improbable within this life-supporting subset of possibilities, but no deeper explanation would be available to us. In effect, some important properties of the Universe, like the value of the cosmological constant, would be random outcomes of the laws of Nature that could take many values consistent with the universal laws. We could no more use the Theory of Everything to determine their observed values than we could use Newton's laws of motion to explain why our solar system contains *nine* planets out to the radius of Pluto.

Many physicists find this a deeply unsatisfactory, even an unacceptable, conclusion. They think that the laws of Nature should offer an explanation for everything. Alas, the Universe was not constructed for our convenience. It would be very suspicious if all our questions about the Universe could be answered *now,* using currently available technology. In other sciences we have got used to recognizing the sorts of questions which a scientific theory cannot answer because of chaotic sensitivity, external inputs of information, or random symmetry breakings. We may have to get used to the same type of indeterminism in our understanding of the properties of the Universe as a whole.

In this situation, we must inevitably weigh up the vital consideration that there is no reason to think our Universe is the 'most likely' with respect to the values of its constants of Nature. On the contrary, we have already seen that our own existence, and that of any conceivable observers, is made possible only by the fact that the values of many of these constants lie very close to those that are observed. Change them a little and there would be no observers. Hence, we see that we must compare the observed values of the constants of Nature not with the most likely set of values arising in the Theory of Everything, but with *the most likely values conditional upon them being such as to allow the future evolution of observers.* And the latter set of values could be very different from the most probable values in the absence of such conditioning (see Figure 5.3). The fact of our own existence, not to say the formidable problem of unravelling all the ways in which these constants of Nature impinge

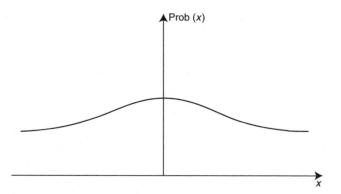

Figure 5.3 A probability distribution in which the probability is not highly peaked about any particular value. If a Theory of Everything predicted that the probability of a constant of Nature, *x*, in our visible universe should look like this, then it is difficult to assess whether we should take the most probable value (zero here) to be the one which should correspond to reality. If some range of possible values of *x* were to give rise to universes which could neither evolve nor sustain intelligent beings, then they could not be observed. We should compare observation with the probability that the constant takes the value *x* conditional on that value permitting the evolution of life, and this can be a very different quantity to the unconditional probability that it take the value of *x*.

upon the cosmological and biochemical conditions necessary for our own or other observers' existence, becomes a vital ingredient in the interpretation of any predictions of the values of the constants of Nature. We need to know all those constants of Nature whose values provide necessary conditions for the existence of observers. This creates an awkward dilemma, for many (if not all) of these constants will be linked to each other through the structure of some Theory of Everything. We need to know that theory completely before we can completely assess what the conditional probabilities for the evolution of complex observers will be. The determination of the constants of Nature now appears to pose a far greater challenge to Theories of Everything than it did before the search for such theories began.

CHAPTER 6

Broken symmetries

Out of this stony rubbish? Son of Man,
You cannot say, or guess, for you know only
A heap of broken images.
—T. S. ELIOT

THE NEVER-ENDING STORY

The search for a Theory of Everything is the search for a universal trivialization—a universal 'nothing but'.
—JEAN-CARLO ROTA

Despite the popular notion that the work of the scientist is discovery—the creation of new ideas and the discovery of new facts about the Universe—many of the books and articles that scientists publish are devoted to a third enterprise: the refinement of existing ideas into simpler, more intuitively embraceable forms, the effacement of the complex into the trivial.

When a new and deep idea is discovered for the first time, it may well appear in a cumbersome language principally designed to express some quite different set of ideas. Gradually, others will re-examine the discovery and find more succinct representations which relate it more naturally to existing ideas. This new relationship may be one of simple logical progression from known ideas or it may be marked by a clash of opposites wherein one must make a radical choice between competing alternatives. This distillation of existing knowledge, to make it simpler and clearer, to refine the true metal of deep truth from the superficial dross surrounding it, is a continuous and vital part of the scientific enterprise. Some scientists are especially good at it and may spend all their activities in furthering its aims rather than pushing forward the far frontiers of discovery.

We witness the consequences of this refinement process in many ways. It acts upon the history of science to distort its progress into artificial channels that lead from an imagined past of ignorance and misconceptions to an enlightened present of right-thinking. It smooths away rough corners and invents motivations. It places disparate individuals in an imagined community of like-minded searchers after truth. Yet it does many of these deceptive things with the best of intentions. Is it not more efficient to teach a subject in a logical rather than a historical fashion? The fact that past investigators took blind alleys is no reason to send others down them before revealing the better path. The result of this historical distillation is undoubtedly to make the laws of Nature appear increasingly simple, compelling, and altogether inevitable to our minds. In recent decades, the discovery of symmetry to be the master key which opens the secret door to the fundamental structures of Nature has been the prime mover in this never-ending quest for an ever more elementary image of things. The longed-for Theory of Everything promises to provide the final discovery after which all physics will become the refinement of its content, the simplification of its explanation. At first, 'the Theory' will be intelligible only to the aficionados, then later to a wider circle of theoretical physicists. Next, it will be presented in ways that make it accessible to scientists dedicated to other disciplines, to students, and then, finally, to educated outsiders and lay-persons. Eventually, it will appear on T-shirts. At all stages of this process, it will be believed that the route from the complicated to the transparently obvious is a path towards the 'true' picture of Nature.*

* This process is something that mathematics shares with the physical sciences. The bulk of the mathematics research literature is devoted to the distillation of known results so that the unintelligible becomes 'obvious' or 'trivial' in the sense that it is merely another manifestation of well-known principles. A case history that is particularly striking is that of the so-called 'Prime Number Theorem' (PNT). The PNT derives from a conjecture of Gauss and Legendre in the eighteenth century and gives a carefully defined approximation to the proportion of numbers below any given value which are *prime numbers* (that is, those, like 5, 7, or 29, that are divisible only by themselves and 1). The theorem was first proved by Jaques Hadamard and Charles de la Vallée-Poussin in 1896 and shows that for all practical purposes there will be virtually no difference between the true fraction of primes and the fraction given by the proposed formula. It states that the fraction of primes to be found amongst the first n integers, when n is large, is approximately equal to the reciprocal of the natural logarithm of n. The first proofs were striking and difficult because they involved complex analysis (a branch of mathematics seemingly far removed from number theory). Then Edmund Landau and Norbert Wiener produced more transparent proofs using simpler concepts. But it was more than thirty years after the original proof before Erdös and Selberg gave, in 1948, what could be termed an 'elementary' proof using the standard ideas conventionally employed in number theory (it was far from simple though, and far from short—it was over fifty pages in length). Twenty years later, in the early 1960s, this proof was refined further to a truly elementary form using undergraduate mathematics by Norman Levinson. Such, more often than not, is the evolution of mathematics. Truly original ideas like Cantor's diagonal argument or Gödel's proof of undecidability, are very rare indeed and are created only a few times a century.

Despite these ongoing tendencies, we are aware of the fact that no matter how often scientists tell us that the laws of Nature are simple, symmetrical, and elegant, the real world isn't. It is messy and complicated. Most of the things that we see are not symmetrical and do not behave in accord with some simple law of Nature. Somehow the breathless world that we witness seems far removed from the timeless laws of Nature which govern the elementary particles and forces of Nature. The reason is clear. We do not observe the laws of Nature: we observe their outcomes. Since these laws find their most efficient representation as mathematical equations, we might say that we see only the solutions of those equations not the equations themselves. This is the secret which reconciles the complexity observed in Nature with the advertised simplicity of her laws. Outcomes are much more complicated than laws; solutions much more subtle than equations. For, although a law of Nature might possess a certain symmetry, this does not mean that all the outcomes of the law need manifest that same symmetry. The fact that our hearts all lie on the left-hand sides of our bodies cannot be taken as any sort of demonstration that the laws of Nature are left-handed.

Another regime in which we learn of the subtle distinction between laws and outcomes, equations and solutions, is at the interface between classical and quantum mechanics. Ever since the invention of the modern formalism of quantum mechanics, we have known how to set about 'quantizing' a particular problem of classical physics so as to extend our understanding of it into the realm of the very small where the act of observership impinges upon the state of the observed. But all this procedure can do is show us how to generate a set of quantum equations (or laws) from the classical ones. There is no known prescription for generating quantum *solutions* directly from classical ones. And indeed no such principle could exist, because there exist solutions (like those describing quantum tunnelling processes) which are intrinsically quantum in character and have no classical counterpart whatsoever.

BROKEN SYMMETRY

Like the ski resort full of girls hunting for husbands and husbands hunting for girls the situation is not as symmetrical as it might seem.
— ALAN MACKAY

The situation in which the outcomes of a law break its symmetry is termed 'symmetry-breaking'. It has been known but not fully appreciated for millennia. And it is responsible for the vast diversity and complexity of the real world.

In Aristotle and his commentators, we find the classic problem of the starving creature caught midway between two stores of food. Buridan's Ass was the most memorable version of this decision problem invented to elucidate the notion that every choice must have a sufficient reason. Leibniz countered it by arguing that the two choices were never identical. There was always some imbalance which led to one choice being taken rather than the other. In modern physics, this question emerges in many situations where an underlying symmetry renders a whole collection of outcomes equally likely. In practice, there must be some particular outcome, and hence the underlying symmetry is broken in the outcome. For example, if a narrow pole is balanced vertically it will fall one way or the other, but this does not mean that the underlying laws of Nature prefer any particular direction in the Universe. A more elaborate example is provided by the phenomenon of magnetization. If a metal bar is heated above a certain temperature, the thermal agitations of its constituent atoms are sufficient to destroy any tendency that they might have possessed to align and define a preferred direction of magnetization. In this hot state, the bar therefore possesses no overall magnetization. But, as the temperature of the bar falls, the thermal agitations decrease in intensity and are unable to randomize the orientations of the atoms. It is no longer energetically favourable for the bar to remain in the state of zero magnetization and it moves towards one of two perfectly symmetrical states as shown in Figure 6.1. These states are characterized by the molecules in the metallic bar aligning themselves in one direction or its opposite. In the first case, we would obtain a bar magnet with North and South Poles as indicated, whilst in the other they would be oriented the other way around. Thus the final state is asymmetric. It has a

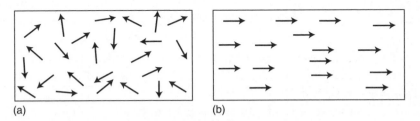

(a)   (b)

Figure 6.1 Magnetization. (a) When the temperature exceeds a critical value, the forces acting upon the atoms in a bar of metal do not distinguish any preferred directions; they are symmetrically distributed because the thermal agitations are large enough to randomize any tendency to point in preferred directions. (b) When the temperature falls below the critical level, it becomes energetically economical for the atoms to orient themselves in the same direction; any direction could be randomly chosen, but, after it has been, the directional symmetry of (a) is broken and North and South Poles are established in the bar magnet.

characteristic orientation. The original symmetry is hidden in the background because *a priori* it is equally probable that the bar be magnetized in the North–South as in the South–North direction.

These examples teach us why science is such a difficult enterprise. We observe the broken symmetries in the particulars of the world and from them we must deduce the hidden symmetries that characterize the laws of Nature.

There is an elegant tapestry of Nature of harmonious weave, but we are seeing the back of it. From the loose threads we must create a picture of the hidden pattern behind it.

NATURAL THEOLOGY: A TALE OF TWO TALES

Sometimes truth comes riding into history on the back of error.
— REINHOLD NIEBUHR

Before we delve a little further into the consequences of this dichotomy between laws and outcomes, it is enlightening to discuss how the history of some natural theology can be most effectively understood by focusing upon this distinction.

Since the Newtonian revolution, there have been two strands to the traditional Design Argument for the existence of God. There have been those like Newton's contemporary apologist, Richard Bentley, who have focused upon the universality and mathematical precision of the laws of Nature themselves to argue for an Author of those laws. Our hymn books bear eloquent witness to the persuasiveness of those

> Laws which never shall be broken
> For their guidance hath He made.

This form of the Design Argument, from the laws of Nature (and sometimes called the eutaxiological Design Argument), appealed most powerfully to physical scientists and astronomers. This was no accident, for these were the scientists whose work impinged most directly upon the pristine symmetries and harmonies exemplified by the laws of Nature. The argument is logically simple, but difficult to appreciate without the benefit of specialist knowledge. Hence it was not so readily taken on board by the interested lay-person. By contrast, there co-existed another form of teleological Design Argument which drew its examples from the marvellous adaptations evident in the natural world. Its staple consisted of apparent contrivances like the human eye and hand, or the way in which the natural environment was tailor-made for the creatures that were to be found in it. This argument, despite being a logical minefield, is graphic and easy to appreciate. Accordingly, it was a powerful persuader of the non-specialist. It held sway amongst naturalists and other keen observers of the minutiae of flora and fauna. This is an argument which focuses upon the outworkings of the laws of Nature—the broken symmetries—rather than upon the laws themselves. It picks upon the innumerable particulars of Nature and points to their correlations with other particulars that are on the face of it totally independent and yet just happen to be harmoniously entwined with them in a singularly appropriate way.

When the Darwinian hypothesis of natural selection was proposed, it provided a general and simple explanation for the contrivances of the natural world—the broken symmetries—but it had no consequences at all for the other form of the Design Argument based upon the laws of Nature themselves. For they are taken to be unchangeable invariants of the Universe. If we study a classic work of natural theology like William Paley's *Natural Theology*, we find both strains of the Design Argument being presented by a writer who had graduated in mathematics but was also a keen naturalist. He presents example after example of the wonderful contrivances of Nature—the eye, the attractiveness of flowers to bees, the apt camouflage of animals—and also extols the fact that the particular law of gravitation that Newton had revealed possesses a multitude of special properties that were necessary conditions for the existence and stability of the solar system and hence of our own existence. Yet, whenever one reads instant criticism of Paley, it cites his book *Natural Theology* only as a paradigm of the former style of (now) naïve Design Argument from special biological adaptation. There is no mention of the second half of his study which deals with the properties of the Newtonian laws of motion and gravitation. Interestingly, Paley himself favoured the biological examples because they are more firmly rooted in observation, and disliked the astronomical examples because there he was divorced from his favourite rhetorical device—analogy. Paley's own expertise and interests were nicely balanced between the two emphases. Although he was trained as a mathematician before entering the Church, he was a keen amateur naturalist.

There is a further strand to this argument that is of interest to us because it provides a link between these two forms of argument that permeated the work of natural theologians during the seventeenth and eighteenth centuries. There existed at that time a strong belief in the notion that some omnipotent Deity controls the behaviour of outcomes that are superficially due to chance because no particular cause can be identified. This control is manifested by the maintenance of stable *average* values of important physical quantities. This notion is of particular interest because it entwines with the early gropings towards the creation of a systematic study of probability and statistics. Newton, as we have said, was the promoter of a Design Argument based upon the precision and universality of the laws of motion and gravitation that he had discovered. But he was also impressed by the peculiar arrangement of the solar system, a feature that he recognized could not be explained using his laws of Nature. A combination of initial conditions and 'chance', which amounts in practice to a collection of chaotic symmetry-breaking processes, was required. In his *Opticks* of 1704, he asks:

> Whence is it that planets move all one and the same way in orbs concentrick, while comets move all manner of ways in orbs very excentric ... blind Fate could never make all the planets move one and the same way in orbs concentrick, some inconsiderable irregularities excepted, which may have risen from the mutual actions of comets and planets upon one another, and which will be apt to increase, till this system wants a reformation. Such a wonderful uniformity in the planetary system must be allowed the effect of choice. And so must the uniformity in the bodies of animals.

William Derham, the author of two extremely successful works of natural theology entitled *Physico-theology* (1713) and *Astro-theology* (1715), wrote to John Conduitt in July of 1733 about a

> peculiar sort of Proof of God which Sir Isaac [Newton] mentioned in some discourse which he and I had soon after I published my Astro-Theology. He said there were 3 things in the Motions of the Heavenly Bodies, that were plain evidences of Omnipotence and wise Counsel. 1. That the Motion imprest upon those Globes was Lateral, or in a Direction perpendicular to their radii, not along them or parallel with them. 2. That the Motions of them tend the same way. 3. That their orbits have all the same inclination.

Newton's recognition of the singular state of the solar system was very influential in motivating later mathematical studies of the *probability* of the solar system's structure arising by chance by Laplace and Bernoulli, and in the whole area of Newtonian science by Abraham de Moivre who dedicated his work *Doctrine of Chances* to Newton, with the stated aim of providing a

> Method of calculating the Effects of Chance ... and thereby fixing certain Rules, for estimating how far some sorts of Events may rather be owing to Design [i.e. determined causes] than Chance ... so as to excite in others a desire of ... learning from your [Newton's] philosophy how to collect, by a just Calculation, the Evidences of exquisite Wisdom and Design, which appear in the Phenomena of Nature throughout the Universe.

Laplace, of course, became famous for explaining all the known motions of solar system using Newton's laws of motion and gravitation, and hence did away with the need for the Deity to intervene to effect periodic 'reformations' of its motions as Newton had proposed. The long-term stability of the average structure of the solar system was thus explained in terms of the properties of the laws of gravitation and motion, although the special arrangement of the solar system remained unexplained in the absence of some detailed theory of how the solar system formed. Later, a similar development influenced the interpretation of many of the special evidences of supposed design in the natural world. Darwin's theory of evolution by natural selection is in practice a statistical theory because we are unable to trace all the causal links in the historical process. Hence the uniformity in the structure of particular species was considered to be a consequence of the stable character of the mean values of such a development, rather than as the consequence of some special choice in the starting state. In 1901, *Biometrika*, the newly formed journal for the statistical study of biological problems, published an editorial justification for its creation which nicely captures the difference in outlook between the physical scientists of the Victorian age, who were preoccupied with the laws of Nature and their exact consequences, and the complicated world of aleatory outcomes that preoccupied the life scientist:

> The problem of evolution is a problem in statistics . . . we must turn to the mathematics of large numbers, to the theory of mass phenomena, to interpret safely our observations . . . The characteristic bent of C. Darwin's mind led him to establish the theory of descent without mathematical conceptions; even so Faraday's mind worked in the case of electro-magnetism. But as every idea of Faraday allows of mathematical definition, and demands mathematical analysis . . ., so every idea of Darwin—variation, natural selection . . .—seems at once to fit itself to mathematical definition and to demand statistical analysis.

THE FLAWS OF NATURE

There is no possibility of reducing all laws to one law . . . no *a priori* means of excluding from the world the unique.
—JOSIAH ROYCE

We cannot expect everything of a Theory of Everything. It may give us all the laws of Nature, but this gift alone does not allow us to explain or derive everything that we observe in the Universe in terms of the principles inherent in the Theory of Everything. Let us investigate the content of this statement in a little more detail.

In our earlier examples of symmetry-breaking, there exists some perfectly symmetrical state which falls out one way or the other because of the breaking of a symmetry. In practice, this will usually mean that some microscopic fluctuation has tipped the scales one way or the other. If this fluctuation is quantum mechanical in origin, then it cannot be traced to a definite local cause and it is thus intrinsically random, rather than merely effectively random because we are simply unable to ascertain its particular cause. Thus symmetry-breaking might be ascribed to random processes at the quantum level. If we were given a bar magnet and asked to explain its structure and behaviour in terms of laws of Nature given by some Theory of Everything, our intuition about things would tell us that there were some aspects of the phenomena that we could not expect to be explained by the Theory of Everything. In particular, we would not try and explain why one end of a magnet was the North Pole rather than the other, or why the magnet was the length it was. Likewise, across the whole range of laboratory physics there are aspects of physical phenomena which are quasi-random as a result of the spontaneous breaking of some symmetry. In solar-system studies, we would not seek to explain why there are a particular number of planets in the solar system. That is too specific a question for a general theory of planetary formation in which many aspects just fall out one way rather than another because of particular starting conditions or random fluctuations. In the situation of the solar system, it is not too difficult to isolate those particulars that one would not seek to explain in terms of the laws of Nature alone. But, when we confront the problems of the large-scale structure of the Universe, it is much more difficult for us to draw the line. Indeed, with our present state of knowledge it is impossible.

The phenomenon of symmetry-breaking introduces an essentially random element into the evolution of the Universe. Certain qualities of the Universe, for example the balance between matter and antimatter, may be determined from place to place by the particular way in which things fell out there. In the laboratory situation, it is usually clear to us which aspects of a physical situation can be attributed to random symmetry-breakings, and so complete explanations for them are not sought in terms of the fundamental laws of Nature. This situation is characteristic of our understanding of the in-between world of condensed forms of matter that is neither of sub-molecular or astronomical scale. By contrast, in a subject like cosmology, we do not yet know which aspects of the large-scale structure of the Universe should be attributed to the laws of Nature and which to the random outworkings of those laws wherein the underlying symmetries are broken. This is a vital distinction to make, because if a feature of the Universe is a consequence of laws or even of initial conditions then there is a case for regarding it as a necessary feature of the Universe that could not have been otherwise. If, on the other hand, the feature is a consequence of symmetry-breaking, then it could have been other-wise and should not be regarded as a key indicator of the structure of Nature. We do not know for example whether the sizes of great clusters of galaxies are inevitable consequences of the laws of Nature, the initial conditions that abided at the Big Bang, or the result of random symmetry-breakings in the early stages of the Universe.

Thus, even if we are in possession of information about the laws of Nature, the initial conditions, and the forces, particles, and constants of Nature, but do not have an understanding of the way in which the symmetries of the laws of Nature and the initial conditions have been disguised by the hierarchy of random symmetry-breakings that occur during the history of the Universe, then our understanding will remain seriously incomplete.

CHAOS

And there was war in heaven.
—THE REVELATION OF ST JOHN

There is a form of symmetry-breaking with a vengeance that has become of considerable topical interest. It is known as 'chaos'. Chaotic phenomena are those whose evolution exhibits extreme sensitivity to the starting state. The slightest change in the starting state results in an enormous difference in the resulting future states. The majority of complicated, messy phenomena, like turbulence or the weather, have this property. The significance of such behaviour was first recognized by James Clerk Maxwell in the second half of the nineteenth century. When asked to lead a conversazione on the problem of free will in his college at Cambridge, he drew his colleagues' attention to systems in which a minute uncertainty in their current state prevents us from accurately predicting their future state. Only if the initial state were known with perfect accuracy (which it cannot be) would the deterministic equations be of use. The neglect of such systems, which are the rule rather than the excep-tion in Nature, had subtly led to a bias in favour of determinism in natural philosophy. The traditional preoccupation with only simple, stable, and insensitive phenomena had created over-confidence in the all-encompassing influence of the laws of Nature. He suggests rather that

> much light may be thrown on some of these questions by the consideration of stability and instability. When the state of things is such that an infinitely small variation of the present state will alter only by an infinitely small quantity the state at some future time, the condition of the system, whether at rest or in motion, is said to be stable; but when an infinitely small variation in the present state may bring about a finite difference in the state of the system in a finite time, the condition of the system is said to be unstable.
>
> It is manifest that the existence of unstable conditions renders impossible the prediction of future events, if our knowledge of the present state is only approximate, and not accurate ... It is a metaphysical doctrine that from the same antecedents follow the same consequents. No one can gainsay this. But it is not of much use in a world like this, in which the same antecedents never again occur, and nothing ever happens twice ... The physical axiom which has a somewhat similar aspect is 'That from like antecedents follow like consequences'. But here we have passed from sameness to likeness, from absolute accuracy to a more or less rough approximation. There are certain classes of phenomena, as I have said, in which a small error in the data only introduces a small error in the result ... The course of events in these cases is stable.
>
> There are other classes of phenomena which are more complicated, and in which cases of instability may occur ...

Maxwell was the first prominent physicist of the post-Newtonian era to focus attention upon the outcomes of the laws of Nature in preference to the forms of the laws of Nature. Newton had built his success upon the recognition of simple general laws which made sense of a vast array of seemingly disparate terrestrial and celestial phenomena. So great was his influence upon the course of scientific development, especially in Britain, that, where the primitive societies had been so preoccupied with the particular events of Nature, the Newtonians were interested only in the lawful aspects of Nature. Newtonianism was more than a scientific method: it was an attitude that permeated all branches of human thinking.

In retrospect, it is curious that it took so long to recognize the extreme sensitivity of many phenomena to their starting state, for there are many walks of life where the effect of a cause is disproportionate and evident. A fascinating early statement of this sensitivity is to be found in Galen's medical writings in the second century AD, where he recognizes the consequences of chance in medical treatment:

In those who are healthy... the body does not alter even from extreme causes; but in old men even the smallest causes produce the greatest change.

In fact, Galen believed that good health was an equilibrium state between two extremes where the 'exact mean of all the extremes' is the same 'in all parts of the body'. Accordingly, he appreciates that it is necessary for any chance deviations from this equilibrium due to external factors to be very small:

Health is a sort of harmony... all harmony is accomplished and manifested in a two-fold fashion, first in coming to perfection... and second in deviating only slightly from this absolute perfection...

The study of chaotic phenomena has proceeded with a methodology that differs significantly from that employed in traditional applications of mathematics to the physical world. In the past, a complicated physical phenomenon like fluid turbulence would have been modelled by attempting to produce as accurate an equation as possible to describe its motions. However, the extreme sensitivity of this type of phenomenon to its initial state also means that it will usually also be extremely sensitive to the form of this equation. If the equation we use contains even the slightest inaccuracy or omission, then very soon the modelled behaviour will deviate dramatically from that which would occur in the real world. As a result of this sensitivity, interest has turned to elucidating the general features shared by almost all possible equations.

Strictly speaking, there can be no common properties of *all* possible equations because any property one cares to list will be manifested by some equation. However, if one refines expectations to the properties of *almost every* equation, that is, excluding only a set of very special cases that are highly unrealistic or improbable, then there do exist general properties common to all the remaining equations. Their discovery has been one of the striking achievements of modern mathematics.

These studies of equations in general rather than equations in particular have revealed to us that chaotic behaviour is the rule rather than the exception. We have come to think of linear, predictable, and simple phenomena as being prevalent in Nature because we are biased towards picking them out for study. They are the easiest to understand. But, we must now swing around to regard it as a mystery that there are such a reasonable number of linear and simple phenomena in Nature. At root, this is why the world is intelligible to us. Simple linear phenomena can be analysed in pieces. The whole is nothing more than the sum of its parts. Thus we can understand something about a system without understanding everything about it. Non-linear chaotic systems are different. They require a knowledge of the whole in order to understand the parts because the whole amounts to more than the mere sum of its parts. Of this we shall have more to say in Chapter 9.

Some of the models which Einstein's theory of gravitation offers as possible descriptions of the very first moments of the Universe's expansion exhibit this chaotic sensitivity to the starting conditions. If, as we discussed in the last chapter in the context of superstring-inspired pictures of its early evolution, the Universe undergoes some transition in which some dimensions of space are confined to an infinitesimally small extent, then the number that escape this fate may be determined in a chaotically sensitive fashion by the conditions that exist in the very early Universe. At the very least they would be expected to vary from place to place. How much of the Universe we are able to deduce from physical or logical principles may therefore hinge precariously upon how delicate is the sensitivity to any initial conditions there may have been.

CHANCE

Statistics is the physics of numbers.
— PERSI DIACONIS

The modern study of chaotic processes regards them as characteristic of the most typical types of change. Only when very particular restrictions are imposed by the situation will insensitivity to starting conditions tend to become the rule rather than the exception. It is the fact that the most general types of continuous change are often those that exhibit a very delicate sensitivity to their precise starting conditions, which results in observable phenomena having very complicated behaviour. A large torrent of turbulent fluid might start with a fairly uniform flow, with different parts of it having very similar speed and direction of motion. Yet, after it falls over the waterfall, these small initial differences between the water motions at neighbouring places become enormously amplified. Before the notion of chaos became well-established, scientists had approached the study of such complicated processes primarily as a statistical problem. That is, they regarded the processes under analysis to be, for all practical purposes, 'random'.

One of the curiosities of history is that the now thriving subjects of probability and statistics did not exist before the mid-seventeenth century. This is the more surprising because when these subjects did appear as part of mathematics they were inspired by the consideration of gambling problems, dice, cards, and all manner of games of chance. Such games have been played the world over for millennia. Indeed, the word 'hazard' derives from the Arabic *al-zahr* meaning 'dice', a game that was played in early Egypt as well as in Greece, Rome, and throughout the Middle East. Why then did the subject of probability—the quantification of chance—not proceed in these cultures as did geometry, arithmetic, and algebra? Unfortunately, there seems to be no compelling answer to this simple question. People were aware of those unpredictabilities that we would call 'chance' and clearly distinguished them from other expected events. They just did not study them as a branch of science or mathematics. More often than not, they used the distinction to proscribe what was lawful and worthy of scientific study.

In some societies, this neglect of chance may be associated with religious beliefs. It is not uncommon for chance phenomena, like casting of lots, to be treated as a means of communicating directly with God (or gods). We recall that the Old Testament Jewish prophet Jonah was thrown overboard after his fellow sailors cast lots and the choice of Mathias by the eleven Apostles to replace Judas Iscariot was made in the same way. Biblical references can also be found to the practice of rhabdomancy, in which sticks are thrown into the air so that omens can be drawn from the directions in which they are found to fall. The Old Testament also speaks on several occasions about the mysterious Urim and Thummim, which were kept in or on the high-priest's linen robe (the 'ephod' as it is called). They were consulted on numerous occasions when national guidance was required and appeared to provide simple 'yes' or 'no' answers, which suggests that they were some form of lot that was either cast or drawn out of a pouch by the high-priest. Another more persuasive proposal is that they consisted of two flat objects, on one side of each was the Urim derived from *'arar* the Hebrew 'to curse'; on the other side of each was the Thummim from *tamam*, meaning 'to be perfect'. Thus, a double Thummim signalled 'yes', a double Urim 'no', and one face of each was read as 'no answer'. In all of these examples, and there are many others like them, the emphasis lies upon using processes which humans cannot foresee as a mouthpiece for the Deity to reveal an unknown but very definite cause. Thus, the casting of lots to identify Jonah as the cause of the storm does not recognize any intrinsically random element in Nature or some mysterious process called 'chance' which operates separately to its usual orderly workings. Rather, it seeks to apportion blame, and discover information that is only available to God in that context because, as Jonah's case illustrates, it was an action against God that the sailors regarded as serious enough to bring about punishment in the form of a storm.

Here, and elsewhere in the Old Testament writings, the introduction of the randomizing process was principally to exclude the possibility of human bias entering the decision. The nearest one gets to any recognition of a notion of 'chance' events unguided by God is in the story of Gideon laying out fleeces on the ground. This he did twice. On the first night he sought dew on the wool but not on the ground, whilst on the second night he looked for the reverse. Presumably, the second trial was to eliminate some possibility that the first sign might have arisen by 'chance', and the change sought in the second state to eliminate what we would now call systematic bias.

These examples lead us to expect that dabbling with random devices was a serious theological business, not something to be trifled with or merely studied for the fun of it. Moreover, the results are not random in any sense that would have been accepted by their witnesses. They were not natural phenomena. Rather, they were the answers of God which were not available to them by other forms of revelation. Thus, although many commentators point to the Old Testament stories as early examples of a well-established familiarity with our modern notion of chance, they are really nothing of the sort.

In other ancient cultures, there is a ready association of randomness with chaos and darkness. These things are undesirable aspects of a dark side of the Universe from which the visible part of it has escaped only by the heroic actions of the gods. Occasional natural phenomena and disasters give us glimpses of this dark side of things. Chance for these thinkers is at root something undesirable, because it is associated with uncertainty and unpre-dictability and hence with danger. If things are not certain—if the harvests and the rains do not come—then there are serious consequences that are readily identifiable with divine punishment.

These more primitive ideas about chance also lead one to expect that it will not become merely another concept for investigation. It is not secular. There will always, of course, be events which because of ignorance appear to be disorderly but which later become accepted into the canon of ordered things when someone notices what the predictable aspect really is. Indeed, this might be viewed as the story of human scientific progress. At first, everything appears random, occult, and attributable to the whims of the gods. As regularities appear, so the personalities of some of the gods must evolve to keep in step with the newly recognized regularities, whilst others are left behind to haunt in the background and provide an explanation for the irregularities. As time goes on, more and more regularities are found and appreciated. They prove to be so beneficial that attention becomes almost exclusively focused upon them and the idea of chance, along with the study of events without discernible causes, simply gets left behind.

Despite religious and social taboos against studying things that happen by 'chance', one might still argue that in many societies there exists such a spectrum of beliefs that there should always be some groups of impious individuals doing bad things like thinking of how they might be sure of winning when gambling.* Many of these people were wealthy and educated. One would have thought that the motivation for a detailed study of particular randomizing devices would have been very great. They would have endowed their owner with a considerable financial advantage over the long run. It is possible that this would simply have been too difficult to do when one recalls that most ancient gambling paraphernalia consisted of objects, like bones or irregularly shaped sticks, which do not possess a number of equally likely outcomes. Indeed, the use of an irregular 'die' of this sort may have been sufficient for its owner to exploit it for profit in the long run simply as a result of experience. By watching its trends over a very large number of trials, it is possible to learn of its biases. By possessing many such devices and playing against a changing roster of opponents, one could exploit that inside knowledge fairly effectively. There is thus no real need for any general theory of such devices; in fact, given that each randomizing device would differ from all others there would appear to be no general theory in any case.

Yet, despite the non-existence of any mathematical theory of chance and randomness in ancient times, there is no lack of appreciation of the general concept of chance and a lively continuing debate as to how one should interpret the existence of events without identifiable causes. One common characterization of the Stoics, which is implicit in the Biblical view that we mentioned above, is to recognize chance as being an anthropomorphism that arises only because of lack of knowledge of the definite but hidden causes of things. This view is expounded in the light of the strictly deterministic views of the Stoics by Cicero, as follows:

> Nothing has happened which was not bound to happen, and, likewise, nothing is going to happen which will not find in nature every efficient cause of its happening...If there were a man whose soul could discern the links that join each cause with every other cause then surely he would never be mistaken in any prediction he might make. For he who knows the causes of future events necessarily knows what every future event will be.

This fatalistic view resonates down the corridors of history, until we recognize its sentiments echoed most forcefully in Laplace's famous passages which

---

* It is interesting to reflect upon the general attitudes towards many forms of gambling in society. It is somehow regarded as undesirable: a vice from which people should be protected. The motivation is not altogether the fear of unpleasant financial consequences since socially acceptable forms of gambling have been created to achieve some general approbation.

introduce his notion of determinism and the ability of an Omniscient Being to foresee the future behaviour of the Universe completely in a deterministic world governed by Newton's laws of motion. For this Intellect, 'nothing would be uncertain'.

The Omniscient Being lurking behind Laplace's thoughts on randomness played a far more important role in another strand of thought about probabilities and chance. The subject of natural theology was an important component of natural philosophy until the mid-nineteenth century. One strand, as we have explained above was concerned with the laws of Nature, the other with the fortuitous outcomes of those laws. The study of both these aspects led to a huge body of opinion in favour of the notion that our Universe was an improbable one. Following Newton, it was not unusual for apologists to consider the unpleasant consequences that would result if the outcomes of the laws of Nature were different, or if the laws themselves were slightly altered. The conclusion invariably drawn from these arguments was that our Universe is extremely improbable, under the implicit assumption that all the alternatives were about equally likely, and hence our particular set of laws and the special outworkings of them required an additional explanation. The explanation most often given was that they had been chosen in order to allow human life to exist. This was a necessary part of the Divine plan.

THE UNPREDICTABILITY OF SEX

The *reductio ad absurdum* is God's favourite argument.
— HOLBROOK JACKSON

A matter that greatly exercised the minds of many natural theologians was the matter of sex. Or, at least, let us say they were interested in a particular aspect of it: the long-term average equality of the birth rates of males and females (in fact the very slight excess of males over females had been noticed as well). A naïve natural subscriber to the teleological Design Argument like William Derham did not ascribe the general regularity or the slight male excess to some long-term mathematical trend, but rather to providential design of a most peculiar sort, because

> surplusage of males is very useful for the supplies of war, the seas and other expenses of the men above the women. That this is the work of divine Providence and not a matter of chance, is well made out by the very laws of chance.

The writer thus appeals to a Design Argument. Later there would arise competing statistical explanations for the facts. Bernoulli showed that the small observed excess of males over females would arise if the probability of a male birth was actually $\frac{18}{35}$ rather than one-half.

None of these natural theological studies involved the concept of chance, which would have been viewed as the antithesis of the design that the hand of God had effected. Chance was personified as a type of cause that symbolized everything that stood in opposition to the established Christian picture of the origin and guidance of the material world. It was not until Maxwell that anyone appreciated the positive attributes of apparently random processes as the generators of many different types of behaviour.*

The natural theologians who studied the living world desired to build into the world every single thing that needed to happen. They did not appreciate that they could endow the world merely with a logic that could respond to every eventuality: it would be Darwin who eventually persuaded us of that. The interesting discovery that we have gradually made over the last century is that random processes possess this responsive capacity. They do not require living systems to be equipped with pre-programming to deal with every possible eventuality. Such a situation would surely render them unviable on the grounds of size and internal complexity alone.

These natural theological arguments about beneficent 'design' to ensure that statistical averages were maintained in a healthy balance had a wide range of supporters, not least of which was the Victorian nursing pioneer Florence Nightingale, who called it 'the thought of God'. They pointed to the equilibrium between births and deaths as well as the balance between the sexes as divinely maintained equilibria. She was much impressed by the stable trends exhibited by individually unpredictable events and was among the first to employ them as a Design Argument for the benign intervention of God. This made her an unexpected student of statistics, of whom one historical commentator wrote:

> [For] her statistics were more than a study, they were indeed her religion. For her Quetelet was the hero as scientist, and the presentation copy of his *Physique sociale* is annotated by her on every page. Florence Nightingale believed—and in

---

* Maxwell's work included the study of the behaviour of molecules in gases where the sheer number of collisions produces an overall situation that defies exact description. Each collision is individually chaotic; yet, because each is effectively independent of the others, a stable statistical pattern of molecular velocities arises. These systems are classic examples of microscopic chaos creating a stable large-scale order. The larger the number of molecules in the system, the smaller will be the occasional fluctuations away from the stable average behaviour.

all the actions of her life acted upon that belief—that the administrator could only be successful if he were guided by statistical knowledge.... Nay, she went further; she held that the universe—including human communities—was evolving in accordance with a divine plan; that it was man's business to endeavour to understand this plan and guide his actions in sympathy with it. But to understand God's thoughts, she held that we must study statistics, for these are the measure of His purpose. Thus the study of statistics was for her a religious duty.

What is interesting about the detailed statistical studies that were made to support these pre-Darwinian apologetics is that they reveal for the first time some belief in mathematical laws governing the development and variation of living things. They recognize that there is a statistical element at the root of reproduction, whose uniformity in the long run requires an explanation. The explanation sought was not a scientific one, but none the less, as with the wider versions of the Design Argument, it played an important role in highlighting key adaptations and equilibria in the natural world that Darwin and his followers were able to focus the search for alternative explanations upon.

SYMMETRY-BREAKING IN THE UNIVERSE

Here lies Martin Englebrodde,
Ha'e mercy on my soul, Lord God,
As I would do were I Lord God,
And thou wert Martin Englebrodde.
— INSCRIPTION ON AN ENGLISH GRAVESTONE

We stressed at the outset of this chapter that the distinction between laws of Nature and their outcomes makes our understanding of the Universe doubly difficult. The overall trend of fundamental physics has revealed that Nature has an observed structure that is strongly temperature-sensitive. When we observe the behaviour of elementary particles of matter at higher and higher energies and temperatures, we find that they become more symmetrical and in some sense simpler. At some very high temperature in the earliest moments of the Big Bang, the Universe could have been maximally symmetric. But, as it expands and the temperature falls, so different options open up for the behaviour of matter and symmetries are successively broken, here in one way, there in another.

Today, we live in a cool low-energy world where biochemistry is possible. Were the Universe significantly hotter or cooler in our locale, we could not have evolved. Thus, we are ill-placed to reconstruct the symmetries of Nature. Of necessity, we live in an era when the deep symmetries of Nature have long since broken and disguised the true simplicity behind things. Moreover, many of the things that we witness in the Universe may have been predominantly determined by the random way in which symmetry-breaking just happened to fall out.

A very specific example of this problem is the idea of the 'inflationary universe', which we introduced in Chapter 3, and which pervades much of modern cosmology. The inflationary universe is not so much a new cosmo-logical theory as an addition to the Big Bang theory of the expanding universe that has been the only acceptable picture of the overall evolution of the Uni-verse since the steady-state theory sunk under a mass of contrary observational evidence in the 1960s. The conventional Big Bang model held to a picture of the Universe in which it expanded from some initial state at a finite time in the past. This expansion is forever decelerating after the start because of the retarding pull of gravity. This property creates a puzzle, because it ensures that the Universe expands relatively slowly in its early stages, so slowly in fact that the huge ball of diameter fifteen billion light years which we call the 'visible universe' today must have expanded from a rather large region in the earliest moments. By 'rather large' we mean that it was vastly larger than any region which light signals would have had enough time to traverse since the expansion began. Hence, the large-scale uniformity of the present-day visible universe, together with the fact that it expands at the same rate in every direction to within at least one part in a thousand, is something of a mystery. The inflationary universe proposes that certain forms of matter, of a sort which particle physicists routinely consider in the mathematical theories they investigate on paper as models of high-energy physics, existed in the first moments of the expansion. The effect of this special form of matter is to make the expansion accelerate for a brief period. As a result, the large imaginary sphere about us that we call the visible universe today can have expanded from a much smaller region during the first moments of the Universe's history. In fact, it easily becomes possible for that embryonic region to be small enough for light signals to traverse it in the time since it began expanding. The pay-off from this speculative investment is that it now becomes readily intelligible why the Universe is so regular over its largest dimensions. The large-scale uniformity is merely a reflection of microscopic uniformity sustained by the

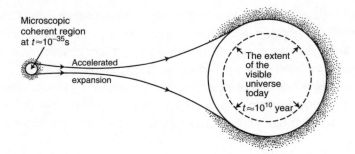

Figure 6.2 The inflationary universe. A schematic representation of the inflationary universe picture, wherein the entire visible universe evolves from the accelerated expansion of a single causally coherent region of microscopic extent. Without the phenomenon of accelerated expansion, these smooth microscopic regions would have evolved into regions far smaller than the size of the observable universe today.

smoothing action of all manner of frictional processes during its first moments (see Figure 6.2).

Let us now imagine the Universe to begin in a fairly chaotic or random state, so that we do not have to make any special assumptions about what its starting state was like. The variation in local conditions will mean that some regions will undergo longer bouts of 'inflation', wherein the expansion accelerates during the early stages of the Universe's history. Symmetry-breaking dictates how physical properties vary from region to region soon after the expansion of the Universe begins. Hence, different regions will inflate by different amounts. Only those which inflate more than some particular amount can live long enough to permit life to evolve. We inhabit one such inflated region. There are many others (an infinite number of them if the Universe is infinite in size) which lie beyond the horizon of our visible universe. These may be very different from the region that we see. Hence, we see why a Theory of Everything leaves us wanting.

The Theory of Everything permits many possible visible universes to exist on the grounds of self-consistency alone, but only the large ones will be seen by physicists. We reside in just one of the possibilities compatible with the necessary conditions for biological evolution. Thus, to understand why our observed part of the Universe possesses the particular properties that it does, we need more than laws of Nature. Many of the striking properties of the large-scale structure of the Universe may be associated with the way in which random symmetry-breakings fell out from region to region. If so, we might look in vain for a direct explanation of those features of the Universe in a Theory of Everything, even if it prescribes the initial conditions at the

Big Bang. The *particulars* of the visible universe will not be derivable from a Theory of Everything if the Universe involves some intrinsically random element, like symmetry-breaking, which can vary from place to place, in its very early stages.

To be more specific about this problem, we can give a specific example that could well turn out to apply in our Universe. One of the most striking properties of the visible universe is the preponderance of matter over anti-matter. Although particle accelerators produce matter and antimatter in equal abundances quite routinely and there is a democratic relationship between the two, we see no antiplanets, no antistars, no antigalaxies, and there is no evidence of any antimatter in the cosmic rays that come from outside our solar system. Nor do we see any evidence of the wholesale annihilation of matter and antimatter which would erupt anywhere in the Universe where the two came into contact. Thus, for some mysterious reason, there exists a form of cosmic favouritism. The observable universe is made of matter rather than antimatter. The other thing that it most obviously consists of is radiation. Indeed, on a straight count the photons have it; for there are on the average about two billion photons of light to be found for every proton in the Universe. Since every time a proton meets an antiproton and annihilates, two photons of light are produced, we can see that a universe such as ours, possessing about two billion photons for every proton, needs to have arisen from a hot dense state in which there were on average a billion and one protons for every billion antiprotons. A billion antiprotons knock out a billion protons producing two billion photons for every left-over proton. But why should the early Universe possess such a weird skewness of matter over antimatter to start off with?

In the early 1980s, a compelling explanation for this cosmic lop-sidedness emerged from the study of unified theories of the strong, electromagnetic, and weak interactions. Their conjunction revealed that an asymmetry should arise naturally because there is a tiny difference in the decay rates of particles and antiparticles in these theories, and this would have an important role to play in the very early stages of the Universe. The final imbalance between protons and antiprotons—the 'billion and one to a billion' bias—can arise from this decay rate asymmetry. The question then remains: what is this asymmetry? In some theories, it is fixed at a constant value that is given by other constants of Nature which we may be able to measure if they determine other observable features of the elementary-particle world. But, in others, this universally constant component is only part of the total asymmetry. It is augmented by another piece that varies randomly from place to place in the Universe because it arises from some random symmetry-breaking process that is sensitive to the local physical conditions of density and temperature. So, in this case, the imbalance between matter and antimatter varies from region to region in the Universe. It is not determined by the 'Theory of Everything' alone. Again, there will be places where the imbalance is so small that lots of annihilation will occur and conditions suitable for the evolution and persistence of life will not arise. Only in those regions where the balance lies between acceptable limits—one such region is obviously our visible portion of the whole Universe—will observers evolve. For those observers, the explanation for the imbalance of matter and antimatter that they see is not to be found in the laws of Nature alone, nor even in the initial conditions. In some sense, there is no explanation of the conventional scientific sort. Conditions fell out in all possible ways in different places. We observe the development of one of the possibilities that allowed life to develop. It is not demanded by the laws of Nature, merely permitted. The fact that it might be a rather improbable one should not worry us either. If the probable outcomes of the Theory of Everything are worlds with similar quantities of matter and antimatter, they will be uninhabited worlds. We would have to find ourselves living in one of the improbable alternatives no matter how great its intrinsic improbability.

We saw in the last chapter that the new developments in our understanding of our prime candidate for a Theory of Everything, string theory, reveals that we might have to recognize many of the Universe's most impressive features as 'merely' the results of symmetry-breaking processes in its very early stages. The process of inflationary expansion that appears to have occurred during the past history of our visible portion of the Universe is expected to result in quite different local properties in other, far distant, parts of the Universe. This means that we may not be able to explain all of the properties of our visible universe in terms of the numerical quantities that define the Theory of Everything. What we see appears increasingly like an outcome—one of infinitely many possibilities—that can emerge from the evolution of the very early Universe.

These examples are but illustrations of how the Universe can be both subtle and malicious. The laws of Nature will not allow us to infer what we will see in the Universe. And we do not even know where to draw the dividing line between those aspects of the Universe which are attributable to law and those which issue from the revolving doors of chance.

What we have learned in this chapter helps us to answer the question 'is the world simple or complicated?' If, like the particle physicist, you look at the world at the level of its laws then it appears very simple. There are a very small number of symmetric patterns, perhaps ultimately just one, which provide the rules of the game for all that happens in the Universe at the level of the fundamental forces of Nature. But, no one has ever seen a law of Nature. We see the outcomes of the laws of Nature, and they are much more complicated, and far less symmetrical, than the laws. All around us we see complicated outcomes of symmetrical laws and the student of this world of outcomes is unlikely to get the idea that the world is simple. This is how it is possible for a Universe like ours to be governed by a very small number of simple laws and yet display an unlimited number of complex states and structures, including you and me.

CHAPTER 7

Organizing principles

Between extremities
Man runs his course.
— W. B. YEATS

WHERE THE WILD THINGS ARE

Doubtless no law of chemistry is broken by the action of the nervous cells, and no law of physics by the pulses of the nervous fibres, but something requires to be added to our sciences in order that we may explain these subtle phenomena.
— WILLIAM JEVONS (1873)

The watcher of science is much impressed by the very large and the very small. The latest speculations about the inner space of elementary particles and the outer space of the astronomer's universe dominate most contemplations of the ultimate stucture of the physical universe. Intuitively, we feel that the ultimate secrets of the Universe's constitution must reside at the extremes of our scales of imagination. But there are extremes other than those of scale and time and temperature. There are extremes of complexity. When we start to tread this new path, we encounter novel and surprising features of the everyday world that reveal the limits of a reductionism that looks to a Theory of Everything to explain the totality of the natural world from the bottom to the top.

In Chapter 5 (Figure 5.1), we displayed a diagram which places the contents of the Universe into an illuminating grand perspective. In Figure 7.1, we have redrawn it to emphasize the hierarchy of complexity in addition to that of size. In the figure are shown the most significant structures in the known Universe. They range from the nucleus of the hydrogen atom, larger atoms, and molecules, to people, trees, and mountains, to asteroids, planets, and stars,

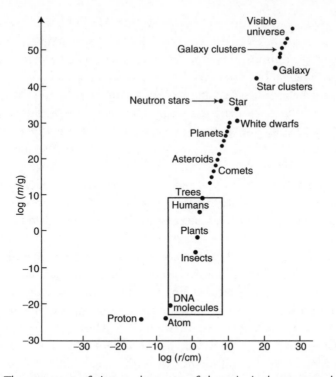

Figure 7.1 The spectrum of sizes and masses of the principal structures known to exist in the Universe (see Figure 5.1), with the region in which complex organized structures form highlighted by the box. Within this region, the phenomena observed are primarily the result of the complex interactions between very large numbers of mutually interacting constituents. They are the results of particular forms of organization between those components, rather than the unusual behaviour of individual forces of Nature alone.

before extending to the largest astronomical structures we see, galaxies and clusters of galaxies, and then to the entire visible universe. There is a 'magic' box that can be drawn on this diagram. It contains the range of structures which exhibit properties characteristic of organized and complex systems. That is, their structure is not simply dictated by the balance between two opposing forces of Nature. Their essential nature is a consequence not of their size alone but of the manner in which their internal constituents are organized. You will notice that human beings sit four-square inside the magic box. This is not surprising. Our brains are the most complicated objects that we have so far encountered in the Universe. We are far from simple. Indeed, were our brains significantly simpler, we would be too simple to know it.

Organized structures inhabit the in-between world betwixt the very large and the very small. Here, we find ourselves in the domain of many-sided complexity: pure and simple. Life is the most conspicuous exhibit in its catalogue of wonders, but not everything that is complicated is alive. From the weather, the behaviour of economies, and opinion polls to exotic materials and unpredictable demographic changes, all possess complexities that defy us to quantify them by our traditional methods. Yet, the very diversity of such organized complexity hints that it might be possible to abstract the very notion of complexity from the specific manifestations of it that we witness, and search for some general principles which govern its emergence and development.

We know from experience that there is a complexity and structure to things that follows simply from there being many of them. The behaviour of a solitary person can be simple; add a second person and entirely new types of complex human behaviour become possible; add a third and even more unusual things can happen; add a dozen more and almost anything can happen. So it is with atoms or electrons. The whole becomes considerably more than the sum of its parts. One reason for this is that, when the number of components becomes very large, it is possible for sub-systems to form. Thus, in the world economy, we do not deal with a single system in which every element is free to interact in all possible ways with all the others. Rather, there are a number of large sub-economies that interact with each other in particular ways.

The motto of the world of complexity is: the more the merrier; the greater the number of components, the greater the number of ways in which they can alter the internal configuration of the system whilst leaving certain gross average characteristics unchanged. As a simple example, consider the number of English words that one can construct from individual letters in a game like *Scrabble*. With a single letter, there is at most one word that can be found in the dictionary. With two letters, one might find a couple. But as the number of letters rises from three to four and beyond, so the number of possible words grows dramatically. And so it is with atoms and particles of matter. The extraordinary phenomena of solid-state physics, like superconductivity and semiconduction, the properties of new materials, all are the result of this complex world of large numbers. Indeed, there has arisen an interesting debate within the scientific community in recent years as to the relative significance of such phenomena compared with those of traditional 'fundamental' science. If one studies the subject matter of popular science books, then it is soon clear to the reader that they are dominated by accounts of black holes, cosmology, and elementary particles. These opposite ends of the size spectrum are perceived as being the most fundamental, the most natural candidates for the tag 'blue-skies' research. To many, the intermezzo world of 'squalid-state' physics is neither as fundamental nor as interesting. Even professional physicists find it to be something of an acquired taste. But to those who have acquired a taste for the study of its complicated natural phenomena we owe an immeasurable debt. For the fruits of their investigation have given us the materials and technologies that underpin the comforts and conveniences of modern living. This, needless to say, has never been in question, but physicists in this area have become anxious not to relinquish the idea that their study is just as fundamental in its own way as that of the cosmologist or the particle physicist. This issue has emerged in several Western societies as decisions have had to be taken about the funding of vast experimental projects to push forward the frontiers of astronomy and elementary-particle physics. The solid-state physicists, the metallurgists, chemists, and material scientists contest that their subjects deserve the same levels of funding by government agencies. The particle physicists claim that their subject is more 'fundamental'. Who is right?

Despite the specific nature of this debate, it is at root an old question dressed up in modern guise. It is the issue of 'reductionism', which has traditionally been primarily of interest to the life scientists. The outright reductionist sees science as a straightforward hierarchy. Starting with zoology, we take the attitude that we 'understand' it when it is reduced to something more basic. For zoology, that something is biology. Biology likewise is founded entirely upon chemistry; chemistry can be shown to be founded upon physics; and physics leads us back to the most elementary particles of matter. When we find them—whether they turn out to be point particles or strings—we have completed the linear chain. Thus, at each stage, the ardent reductionist claims that there is a 'why' question that always points in the same direction: inwards to the smaller scale. The workings of the magic box are always to be found on the inside. Nobel-prizewinning particle physicist Steven Weinberg articulates this view as part of his argument for the funding of a future particle accelerator in the face of criticism from many condensed-matter physicists:

> Still, relying on this intuitive idea that different scientific generalizations explain others, we have a sense of direction in science. There are arrows of scientific explanation, that thread through the space of all scientific generalizations. Having discovered many of these arrows, we can now look at the pattern that has emerged, and we notice a remarkable thing: perhaps the greatest scientific discovery of all. These arrows seem to converge to a common source! Start anywhere in science and, like an unpleasant child, keep asking 'Why?' You will eventually get down to the level of the very small ... I have remarked that the arrows of explanation seem

to converge to a common source, and in our work on elementary particle physics we think we're approaching that source. There is one clue in today's elementary particle physics that we are not only at the deepest level we can get right now, but we are at a level which is in fact in absolute terms quite deep, perhaps close to the final source.

Here we see the argument for fundamentality point towards the discovery of a Theory of Everything. It is believed that there is a Theory of Everything and we are close to finding it; for Weinberg concludes:

> There is reason to believe that in elementary particle physics we are learning something about the logical structure of the Universe at a very very deep level. The reason I say this is because as we have been going to higher and higher energies and as we have been studying structures that are smaller and smaller we have found that the laws, the physical principles that describe what we learn become simpler and simpler...the rules that we have discovered become increasingly coherent and universal. We are beginning to suspect that this isn't an accident, that it isn't just an accident of the particular problems that we have chosen to study at this moment in the history of physics but there is a simplicity, a beauty, that we are finding in the rules that govern matter that mirrors something that is built into the logical structure of the Universe at a very deep level.

There is clearly something in what the ardent reductionist has to say. There is no reason to believe that the stuff of biology is made of anything but the atoms and molecules that the chemist studies; nor any reason to think that those atoms and molecules are composed of anything but the elementary particles of the physicist, any more than we would doubt that Michelangelo's *Pietà* is composed of raw material other than marble and stone. But such reductionism is trivial. It was worth stating only when there were baseless speculations that some mysterious substance ('phlogiston') was present in fire or some *élan vital* in 'living' things. As we bring simple things together, they produce aggregates that exhibit a wider diversity of behaviour than the sum of their parts. Thus qualitatively new phenomena appear as the level of complexity rises or the number of ingredients increases. Such a situation was not foreseen by the early vitalists. As C. H. Waddington remarked:

> Vitalism amounted to the assertion that living things do not behave as though they were nothing but mechanisms constructed of mere material components; but this presupposes that one knows what mere material components are and what kind of mechanisms they can be built into.

A collection of 10^{27} protons, neutrons, and electrons may be all that a desktop computer *is* at some level, but clearly the way in which those sub-atomic particles are put together, the way in which they are organized, is what distinguishes the computer from a crowd of 10^{27} separate sub-atomic particles. Thus, at this level, that of the possible behaviours that the system can manifest, the computer is more than the sum of its parts and what makes it so is the way in which the atoms are bonded together to form particular types of material and the way in which those materials are hard-wired together into switches and circuits. The properties of the computer are a manifestation of a particular level and quality of complexity being attained. The larger and more complex the internal circuitry and logic, so the more sopisticated will be the capabilities of the device.

These examples teach us that if reductionism means that all explanations for complexity must be sought at a lower level, and ultimately in the world of the most elementary constituents of matter, then reductionism is false. Instead, we might expect to find novel types of complex organization, at each level, as we go from the realm of quarks to nucleons to atoms to molecules to aggregates of matter. Each of these new behaviours will be essentially a manifestation of a particular level of organization having been attained under particular environmental conditions. One of the most striking things that distinguishes such complex phenomena from those simpler phenomena of interest to the elementary-particle physicist is that the latter are believed to be fully exhibited at some stage in the history of the Universe. If we follow the Universe backwards far enough in time, we should encounter natural conditions extreme enough to produce all the most elementary particles of Nature in a free state. They do not require anything special about the environment except high temperature. But the outworkings of complexity are altogether different. They are usually very sensitive to many of the details of the environment and do not arise 'naturally'. That is, we often have to engineer the special conditions under which the phenomenon will appear. It is quite easy to imagine that some of the complex phenomena that we have been able to produce in the laboratory or the factory have never before been manifested in the Universe. This is a sobering thought: a property of matter, like high-temperature superconductivity, may never have occurred naturally during the entire history of the Universe. It was latent in the laws of Nature, but can be exhibited only when very particular artificial conditions are met, and only then when matter is organized in a peculiar and 'unnatural' manner.

Life as we know, and partially understand it, is a classic example of what can occur when a sufficient level of complexity is attained. Consciousness appears to be a manifestation of an even more elaborate level of organization.

Accordingly, both of these phenomena are very sensitively attuned to the environmental conditions within which they are found. This is hardly surprising if they are the products of a process of natural selection in which the environment plays a key role in determining the nature of those advantageous attributes which will be selected for in subsequent generations. Yet it is surprising to find that the form of life that we know, and are, is very finely balanced with respect to the astronomical environment and even to the forms of the laws and constants of physics.

Biologists have agreed upon no general definition of life. Our experience of its possible forms is too limited (some of the known examples are shown in Figure 7.2). None the less, while there is a lack of agreement as to what properties are *necessary* for something to be called 'living', there is a reasonable consensus about those features which would be *sufficient* for something to be termed 'living'. Any attempt to spell out necessary conditions tends to degenerate towards a specification that is very narrow, a specification that is little more than a description of known forms of life. It is most useful to propose that a sufficient condition for something to be termed living is that it can reproduce itself in some environment and must contain some level of organization which is preserved by natural selection. Reproduction does not mean that exact copies are made in each generation, merely that an exact copy would have a higher probability of survival than close copies in the same environment. In any biosphere, some, although technically not all, organisms would satisfy this definition. For instance, whereas a single human being fails to satisfy it (it cannot reproduce alone), it is composed of many cells which satisfy the definition. A male and a female would together satisfy the sufficient condition as well.

The complexity of life as we know it has made it a rather parochial affair. There is no evidence of any other form of complexity worthy of the name of 'life' in our solar system, where we have searched as well as listened, or in our Galaxy, where we can only listen. The latter silence tells us that certain species of complexity, those that are advanced enough to launch spaceprobes or send radio messages, either do not exist or do not wish to communicate. The absence of that level of organized complexity in the solar system is not entirely surprising: complexity is a delicate business. Chemical and molecular bonds require a particular range of temperature in which to operate. Liquid water exists over a mere one hundred degree range on the centigrade scale. Even Earth-based life is concentrated towards particular climatic zones. The temperature at the Earth's surface keeps it tantalizingly balanced between recurrent ice ages and the roasting that results from a runaway greenhouse

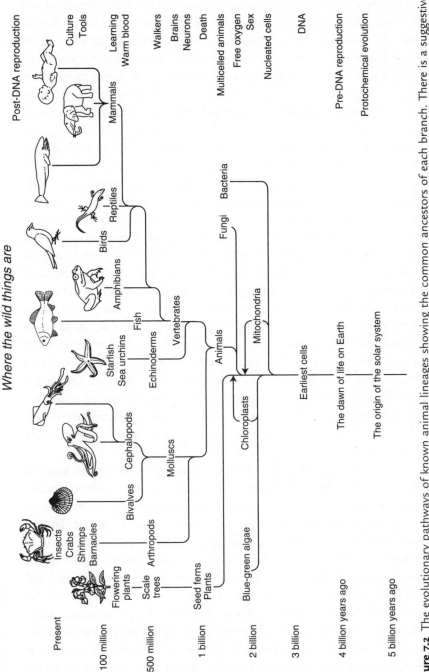

Figure 7.2 The evolutionary pathways of known animal lineages showing the common ancestors of each branch. There is a suggestive parallel between the mobility of each life-form and its intelligence. Humans are distinguished further by the highly effective way in which they have pooled the individual intelligence of single individuals to produce a collective intelligence that greatly outweighs the capability of any single individual.

effect. Very slight differences in the size of our planet or its distance from the Sun would have tipped the scales irretrievably towards one or other of these fates. That such a delicate balance, which is essentially the outcome of those random symmetry-breakings that we discussed in Chapter 6, should be so crucial suggests that natural complexity may be a rather rare thing in the Universe.

The most elaborate complex constructions that the laws of Nature allow all require intermediate steps for their natural attainment. At present, we are one of those intermediate steps. Biochemists believe that, whereas we can envisage different forms of life, based upon chemistries other than carbon or even based upon something non-chemical, only carbon-based life can evolve spontaneously. Other forms of complexity deserving the name 'life' can only come into being non-spontaneously with the aid of the complex operations that can be carried out by carbon-based life. To give a simple illustration, we might consider the computer revolution that has taken place in the West during the last decade. This is an evolutionary process. Generations of small computers are 'reproduced' by manufacturing processes, each is an improvement upon the previous model by virtue of some information fed in from users or the market. Those brands that are defective or inferior gradually become extinct or are subsumed within others. This form of evolving complexity is based upon silicon rather than carbon. Science fiction writers have long realized that the element silicon (the most abundant material in the Earth's crust) possesses, in a markedly less spectacular way, some of the unusual stability, flexibility, and bonding properties of carbon atoms which allow it to form the long chain molecules that are the basis of organic chemistry. Although silicon does have a very limited capacity to form chain molecules, it tends to create solid crystal lattices like quartz (silicon dioxide) rather than liquids and gases or complicated reactive chain molecules. However, silicon and related elements have collective properties that have made them the basis for the microelectronics and computer industries. Today, a science fiction writer looking for a futuristic tale of silicon dominance would not pick upon the *chemistry* of silicon so much as the *physics* of silicon for his prognostications. But this form of silicon life could not have evolved spontaneously: it requires a carbon-based life-form to act as a catalyst. We are that catalyst.

A future world of computer circuits, getting smaller and smaller yet faster and faster, is a plausible future 'life-form', more technically competent than our own. The smaller a circuit can be made, the smaller are the regions over which voltages appear, and hence the smaller these voltages can be. Tiny layers of material just a few atoms thick allow the electronic properties of a material to be finely tuned and rendered far more effective. The first transistors were made of germanium, but were far from reliable and failed at high temperatures. When high-quality silicon crystals could be grown, they were used in a generation of faster and more reliable silicon transistors and integrated circuitry. Newer materials like gallium arsenide allow electrons to travel through them even faster than through silicon and has given rise to the line of Cray supercomputers. The evolution of computer power is represented in Figure 7.3. Undoubtedly other materials will eventually take over. The story may even come full circle back to carbon again. Pure carbon in the form of diamond is about the best conductor of heat, a property that is a premium in a densely packed array of circuits.

BIG AL

A man, viewed as a behaving system, is quite simple. The apparent complexity of his behaviour over time is largely a reflection of the complexity of the environment in which he finds himself.
— HERBERT SIMON

If we take the short-term position that all forms of life and extreme complexity other than those that are carbon based cannot evolve spontaneously in the time available to them since the first stars and planets formed, then we can classify all these other forms of complexity under the heading of 'artificial life' (AL). This subject should be compared, but not confused, with the study of artificial intelligence (AI): we are interested in a broader spread of complex processes than those which mimic cognitive processes. One of the workers in this field has described its most optimistic goal as the desire 'to build models that are so lifelike that they would cease to be models of life and become examples of life themselves'. In practice, this amounts to the study of all forms of organized complexity with special emphasis upon those varieties that change in time and interact with their environment. Even without the added subtlety of input from a changing environment, one can demonstrate rather interesting general results which illustrate what is possible in principle when the form of artificial life (or complexity) is constructed with particular properties. For example, one can imagine a deterministic form of artificial life which, once set in operation, requires no additional control or input but will reproduce itself indefinitely, creating a sequence of progeny, each one of

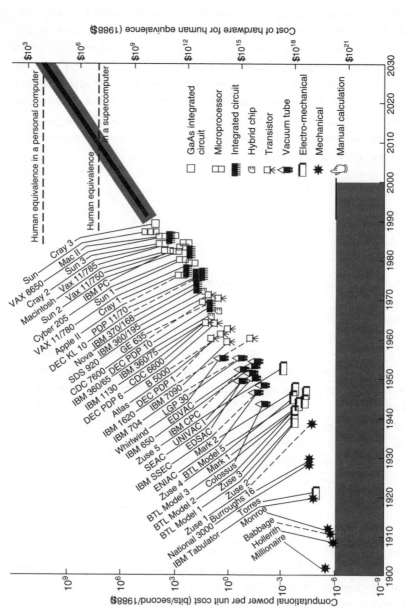

Figure 7.3 The evolution of computer power during the twentieth century. Also shown is the equivalent measure of unaided human calculating power and the technological evolution of the computer industry from mechanical devices through electrical machines to contemporary electronic processors.

which is superior to its parent. We can imagine an artificial life-form which has information storage capacity able to embody all the axioms and rules of arithmetic. It can therefore generate theorems of arithmetic. The sum total of all these theorems could be defined as its 'intelligence'. However, our earlier discussion of Gödel's incompleteness theorem tells us that the intelligence of the life-form cannot include all the truths of arithmetic. There must always exist some that it can neither prove not disprove. But when the organism discovers (as it can) that some statement of arithmetic is neither provable nor disprovable from the set of axioms that it has embodied within it, then it can simply add the undecidable statement as a new embodied axiom. The enlarged axiomatic system must still be incomplete in some new way, of course, but the organism now evolves by repeating this procedure: identifying undecidable propositions before incorporating them as new axioms, getting smarter and smarter because each progeny can prove all the theorems that its parent could (some of them by far shorter sequences of logical deduction because of the extra power derived from its additional axioms which therefore permit new sequences of logical deduction) plus some new ones because of its extra axiom. The information content of each offspring exceeds that of its parent. A further twist could be added here by having two offspring born from each parent: one would incorporate the chosen undecidable proposition as a new axiom, whilst its 'brother' would incorporate its negation as an axiom.

The primary features that characterize the deductive ability of any form of organized complexity are the rate at which it can process information (that is, transform one set of numbers into a new one) and the size of its memory store. Memory size dictates the ability of a system to learn and adapt to change. In Figure 7.4, we can see a comparison of these two attributes of a wide range of complex systems, some of which we would regard as living, others which we would not.

This vague distinction is one that we tend to make on the basis that, somehow, living things are always wet and soft but non-living things tend to be hard and metallic. Computers and crystals do not look like other forms of life. But this is a rather subjective distinction, especially when we look back at the sequence of events that may have led to the evolution of carbon-based 'wet-ware' that forms our existing flora and fauna.

Graham Cairns-Smith of the University of Glasgow has proposed that the natural form of life that we see now may not have been the primary source of the complexity based upon carbon chemistry that characterizes current living organisms. In his scenario of 'genetic takeover', he suggests that the first

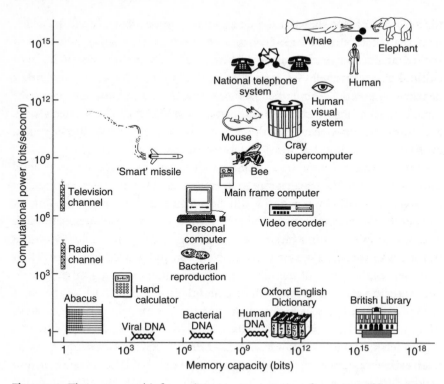

Figure 7.4 The power and information storage capacity of a variety of living organisms and human technological creations.

'organisms' were tiny crystals of clay* which changed by the familiar processes of fracture and crystal growth.

The pattern of the crystal structure contains various irregular patterns called 'defects'. These defects play an important role in the history of the clay, because they affect its physical and chemical properties, so altering its effectiveness as a catalyst in chemical reactions with neighbouring substances. Eventually, Cairns-Smith suggests, some of the crystals randomly incorporate the ability of adjacent carbon compounds to do more complicated things, store patterns, and ultimately produce molecules that could produce replicas of themselves. Once this process begins, the crystal basis is rapidly taken over by the far more efficient carbon machinery. The evolutionary result will be carbon-based life-forms with little or no trace of their vestigial crystalline origins. This whole process of genetic takeover is rather similar in style to

* It is a little known historical fact that the Reformation theologian Martin Luther was the first to consider the possibility of life based upon clay. He writes that, 'If God consulted me I should have advised Him to continue the generation of the species by fashioning them of clay'.

the takeover of the United Kingdom's motor car industry by the Japanese, or perhaps the future silicon take-over of our own carbon-based chemistry. Indeed, at a deeper level we can detect its hand in most of the intellectual and cultural trends in which we participate. When someone has a new idea, it will be taken on board by some other innovator who, at first, thinks about it in the same context as its originator, but then will sense the scope for improvement and transplant the essence of the idea into another context. The idea has evolved. It has been taken over by a new mind.

Our digression into things living is demanded by the emphasis placed upon both the understanding and the simulation of 'life' by so many modern investigations. The entire panoply of such studies now falls within the embrace of 'cognitive science'. At root, such investigations are faced with understanding a particular type of complexity, albeit one that is dauntingly multifaceted. Yet our focus upon living systems was, in this case, motivated solely by the fact that they are the most complex things that we see, rather than by any desire to endow them with some supernatural significance.

We have seen that a naïve reductionism that would seek to reduce everything to its smallest constituent pieces is misplaced. If we are to arrive at a full understanding of complex systems, especially those that result from the haphazard workings of natural selection, then we shall need more than current candidates for the title 'Theory of Everything' have to offer. We need to discover if there are general principles that govern the development of complexity in general which can be applied to a variety of different situations without becoming embroiled in their peculiarities. Perhaps there exist a whole set of basic rules about the development of complexity which reduce to some of our simpler laws of Nature in situations where the level of complexity is essentially nil? If such rules do exist, then they are not like the laws which the particle physicists seek. But is there any evidence that such principles might exist?

TIME

If everything on Earth were rational, nothing would happen.
— FYODOR DOSTOEVSKY

The nature of time is one of those baffling problems that physicists have debated for centuries, but have made depressingly meagre progress in unravelling. New scientific theories, be they relativity or the quantum theory, invariably bring with them a new perspective upon the nature of time, but they usually add another puzzling aspect to lay alongside those we have already, rather than present a decisive new viewpoint that replaces all that went before. Our discussion of organizing principles provides a natural juncture at which to highlight a historical tension in the attitude of thinkers towards the nature of time. Opinion has ebbed and flowed between two extreme views for thousands of years, but the investigations of complex and organized systems may signal a turning of the tide towards the extreme that has been out of favour for most of the twentieth century.

From the days of the earliest Greek thinkers, there has existed a dichotomy between those who were willing to recognize the role of time in natural processes as an essential feature in the actuality of the world. These thinkers, like Aristotle and Heraclitus, placed emphasis upon the observed 'world of happenings' as the true reality to which all attempts at explanation and enquiry should be addressed. In sharp contrast to this pragmatic approach, there has always existed a tradition, starting with Parmenides and later matured into its most elegant espousal by Plato, that we should attempt to eliminate time from our picture of reality. It should be hidden or reduced to something else. Plato effected this effacement by attributing ultimate significance to other-worldly forms which provided the perfect blueprints from which all observed phenomena derived, albeit imperfectly. These eternal forms were timeless invariants, the true reality of which observed things were but imperfect shadows. Here we see the de-emphasis of the role of time. The ultimate things were not changing in time. Only the imperfect approximations to them displayed variability, and hence it is easy to discount time as being not of the true essence of things. This bias we see displayed explicitly in the body of early Greek mathematics and science. They were interested in what we would now call statics: perfect circles, invariant harmonies, the meanings of pure numbers. Platonic idealism has a natural tendency to ascribe some form of unchangeability to the ultimate realities.

Newton and the scientists who followed in his footsteps were not primarily interested in static harmonies. For them, laws of Nature meant laws of change—dynamics. Time had an explicit role to play. But that did not shed any light on what it was. To avoid becoming embroiled in 'hypotheses', Newton wrote in the early pages of the *Principia*:

> I do not define time, space, place and motion, as being well known to all. Only I must observe, that the common people conceive those quantities under no other notions but from the relation they bear to sensible objects.

His course of action was to erect time as a fixed external standard that was unaffected by any events that occurred in the Universe. This distinguished it from the 'common' notion that he refers to which always associates the passage of time with some sequence of events (like the movement of the Sun across the sky) and thereby attributes some aspect of a temporal nature to those objects.

In the post-Newtonian era, there was to emerge a perspective that grew in influence and came to dominate the view of the world taken by most physicists until comparatively recently. It was discovered that there exist certain conserved quantities in Nature, like the total energy or momentum involved in an isolated process. So, despite the superficial appearance of change in some complicated natural processes, there exists an unchanging underlying aspect that reflects an invariance of the laws of Nature. Consequently, it is possible to represent all traditional laws of change governing motion by equivalent statements that certain quantities remain *invariant*. Here we see the Platonic strand re-emerge. Time is de-emphasized and the invariance of certain things is taken to be more fundamental than the rules governing the allowed changes in time that are permitted by these invariances.

From the early 1970s until just a few years ago, this approach underpinned the dramatic progress made by elementary-particle physicists through the formulation of *gauge theories*, which we introduced in Chapter 4. They derived the laws governing changes in transmutations of, and interactions between, elementary particles by the primary assumption of an invariance of things with respect to certain classes of changes in space and time. The great success of this approach reinforced the general tendency to place greatest significance upon the timeless aspects of reality: the conserved quantities of Nature and their associated symmetries, equilibria, and invariances. Only in the last decade has this emphasis ceased to be the dominant one in the physical sciences. There has grown up a renewed interest in the particular rather than the general. This, as we have seen in our earlier discussion of symmetry-breaking, has been brought about by a recognition of the extraordinary richness displayed by the outcomes of laws of Nature that is not shared by the laws themselves. This study of outcomes has focused upon the evolution of complex systems, symmetry-breaking, and chaotic behaviour. In all these things, time is of the essence. Invariance plays a weak role that sheds little or no light upon the essential properties of the phenomena in question. There is a fundamental reason why many such phenomena must be set in opposition to the search for time invariance in Nature. When we encounter sequences of events which display behaviour that is algorithmically incompressible, this means that they admit of no abbreviated description. They cannot be encapsulated into some simple formula that contains the same information content. In particular, this means that an algorithmically incompressible process cannot be replaced by some invariance principle. It is its own simplest representation, and hence the entire sequence is required to describe it. Thus we see here the Aristotelian emphasis upon events and the relation between events in time re-emerging as a dominant consideration in the description of the natural world in opposition to the focus upon invariance. When we look at the elementary-particle world, we see invariance as a shining beacon to guide us into the ways of the world; when we peer into the middle ground, where complexity and organization dictate the structures that exist, we find that time and change are essential features of the fabric of the world.

BEING AND BECOMING ORGANIZED

The Three Laws of Robotics

1. A robot may not injure a human being, or, through inaction, allow a human being to come to harm.
2. A robot must obey orders given it by human beings, except where such orders would conflict with the First Law.
3. A robot must protect its own existence as long as such protection does not conflict with the First or Second Law.

— ISAAC ASIMOV

Organizing principles are likely to differ from conventional laws of Nature because they would need to apply to systems of a finite size. They will not dictate how elementary particles move. Rather, they will constrain how an entire collection of things can be configured. An example which is familiar is the so-called second law of thermodynamics, which governs the behaviour of whole volumes of things. In simple language, it requires that the degree of disorder (which can be defined precisely) in a closed system can never decrease as time passes. This tendency, so evident in many aspects of things, has been of recurrent fascination to thinkers in all subjects. It is no accident that it emerged as a fully-fledged branch of science in the second half of the nineteenth century, during the heart of the industrial revolution. The study of steam engines led not only to an understanding of the degradation of energy from useful ordered forms into useless disordered forms but also to the paradigm of the Universe as a vast engine degenerating slowly into a cosmic heat death. This notion generated a curious philosophic pessimism during the early decades of the twentieth century and it became rather fashionable in literary circles to know about the second law of thermodynamics. One recalls that C. P. Snow used it as a touchstone for the scientific literacy of non-scientists:

ignorance of it was tantamount to a scientist not having heard of Shakespeare. We shall have something to say about the modern analysis of this particular problem in due course, but here we wish to stress the *universality* of the second law of thermodynamics. This is a feature that must be shared by any principle laying claim to govern the universal development of complexity.

The second law of thermodynamics is seen to govern the behaviour of heat engines and chemical reactions: this much we would expect. But in the mid-1970s a rather unexpected discovery was made which surprised physicists and renewed their confidence in the second law of thermodynamics as a guiding principle in areas of science far removed from those that played midwife at its conception and where far more complicated concepts might have been suspected to play a guiding role. In the early 1970s, astrophysicists were preoccupied with their first detailed discoveries about the structure of black holes. Black holes are the simplest objects in the Universe. They are created when a large quantity of mass is attracted by the force of gravity into a sufficiently small volume of space. The strength of the resulting gravity field ensures that there arises an imaginary surface, or *horizon*, around it and no particle or light signal from within this horizon surface can pass to the outside. The black hole contains the material within this surface, but it is not a solid object. Although material within the horizon will continue to fall towards its centre and become involved in all manner of complicated antics, none of this is visible to any outside observer. All that he can determine about the matter within the horizon is its total mass, together with any net electric charge or angular momentum (a measure of the overall rotation) that it might possess. These are the only things that can be known about a black hole; this makes them the simplest objects in the Universe. Other objects, like stars or people, require countless quantities to be known in order to specify them uniquely. The three defining quantities of a black hole are not unexpected: they are the ones that have been found to be absolutely conserved in all observed physical processes in the Universe. The fact that they are properties of black holes guarantees that they can continue to be conserved in Nature even when black holes are present. What is most interesting about this state of affairs is the enormous list of things that are not available to the outside observer once a complicated configuration of matter becomes enclosed within its horizon. The outsider cannot tell whether the inside of a black hole contains matter or antimatter, positrons or protons, brass bedsteads or the works of Proust. The information that makes such distinctions does not penetrate the horizon.

The most general possible black hole that Einstein's theory of gravitation allows was found in the early 1960s, and physicists then set about trying to understand how changes can occur when matter is added to a black hole or when two or more black holes merge to form a new enlarged black hole. A number of simple rules were found to govern any processes involving black holes and other forms of matter. The *gravitational field* must have a constant strength all around the horizon of a black hole. The total *surface area* of all the horizon surfaces of the participating black holes can never decrease. Any changes in a black hole's mass, electric charge, or angular momentum are linked together in a definite manner. Three laws governing black hole changes were thus found, but it was soon noticed that something unusual was going on. If one merely replaced the words 'surface area' by 'entropy' and 'gravitational field' by 'temperature', then the laws of black hole changes became merely statements of the laws of thermodynamics. The rule that the horizon surface areas can never decrease in physical processes becomes the second law of thermodynamics that the entropy can never decrease; the constancy of the gravitational field around the horizon is the so-called zeroth law of thermodynamics that the temperature must be the same everywhere in a state of thermal equilibrium. The rule linking allowed changes in the defining quantities of the black hole just becomes the first law of thermodynamics, which is more commonly known as the conservation of energy.

At first, this surprising concurrence was regarded as something of a coincidence. Black holes, by definition, could not have any temperature other than zero. Nothing could escape from their surface, so their radiant energy must be zero to any outside observer. Put a black hole in a box along with heat radiation at some fixed temperature and the two do not come into equilibrium at some new temperature. The black hole just gobbles up all the radiation.

For these reasons, the thermodynamic analogy was regarded by many physicists as little more than a curiosity. After all, it was not imagined that thermodynamics would have anything to do with the laws of gravitation that apply to the strong gravitational fields at the horizon surface of black holes. What could be less like a steam engine? Then, in 1974, Stephen Hawking made a dramatic discovery. He decided to examine for the first time what occurs when one applies the notions of quantum mechanics to black holes. What he discovered was that black holes are not completely black. When quantum mechanics is included in the discussion of their properties, it is possible for energy to escape from the surface of the black hole and be recorded by an outside observer. The variation in the strength of the gravitational field near the horizon surface is strong enough to create pairs of particles and antiparticles spontaneously. The energy necessary to do this is extracted from the source of the gravitational field, and, as the process continues, so the mass of the black hole ebbs away. If one waits long enough, it should disappear completely unless some unknown physics intervenes in the final stages. Such a discovery was exciting enough, but its most satisfying aspect was the fact that the particles radiated away from the surface of the black hole were found to have all the characteristics of heat radiation, with a temperature precisely equal to the gravitational field at the horizon and an entropy given by its surface area, just as the analogy had suggested. Black holes did possess a non-zero temperature and obeyed the laws of thermodynamics, but only when quantum mechanics was included in their description.

The deep significance of this discovery appears to be that we have found a physical situation where two different natural principles, of quantum mechanics and general relativity, come together, which admits of a simple thermodynamic description. We expected all the rules governing how things behave in such a quantum gravitational situation to be complicated and novel. Many undoubtedly are; yet we find that the tried and tested principles of thermodynamics encompass them within their dominion. Besides giving physicists confidence that they might be able to elucidate still more complicated problems of basic science by appeal to simple thermodynamic principles, this case history bolsters our faith in thermodynamics as a paradigm for a 'law' governing the organization of complex systems.

At first, one might think that something like thermodynamics is a rather restrictive concept because it concerns itself with temperature and heat. But its application is not just restricted to all things thermal. It is possible to relate the notion of entropy, which is a measure of disorder, to the more general and fruitful notions of 'information', of which we have already made use in discussing the richness of certain systems of axioms and rules of reasoning. We can think of the entropy of a large object like a black hole as being equal to the number of different ways in which its most elementary constituents can be rearranged in order to give the same large-scale state. This tells us the number of binary digits ('bits') that are needed to specify in every detail the internal configuration of the constituents out of which the black hole is composed. Moreover, we can also appreciate that, when a black hole horizon forms, a certain amount of information is forever lost to an outside observer. The area of the horizon—the entropy of the black hole—is then intimately related to the quantity of information lost to the outside observer when a horizon forms around a region of the Universe to create a black hole.

The success of discovering a thermodynamic principle associated with the gravitational field of a black hole has led to a speculation that there might exist some thermodynamic aspect to the gravitational field of the whole Universe. The simplest assumption to make, following the black hole case, would be that it is the surface area of the boundary of the visible universe. As the Universe expands, this boundary increases and the information available to us about the Universe increases. But this does not seem promising. It would appear to tell us only that the Universe must continue expanding forever, for if it were ever to begin to recollapse the entropy would fall and violate the second law of thermodynamics. The Universe can expand in all sorts of different ways and still have increasing area. What we really want is some principle that tells us why the organization of the Universe changes in the way that it does: why it now expands so uniformly and isotropically.

One interesting development that has emerged from the study of black holes and information is a new fundamental principle governing the maximum information content of a volume of space. One might have thought that this would be proportional to the volume itself, since this would limit the mass of information storage memory. However, it appears that the maximum information content is determined by the surface area of the volume, just as for a black hole. More interesting still, the maximum information content within a bounding surface area corresponds to the information, or entropy, that results if it is the surface of a black hole of the same volume. This 'holographic principle', as it has become known, elevates the surface of regions to a special status. When it comes to the visible universe the situation could be subtle. The three-dimensional volume of space might be the surface area of a four-dimensional volume.

These tantalizing connections between the maximum amount of information that can be stored in a region of space and the theory of black holes and their thermodynamics, has maintained a hope amongst physicists that there might be a simple thermodynamic interpretation of a Theory of Everything that could cut through all the mathematical complexities of string theory and the search for an underlying M theory.

THE ARROW OF TIME

Time travels in divers paces with divers persons. I'll tell you who Time ambles withal, who Time trots withal, who Time gallops withal, and who he stands still withal.
— WILLIAM SHAKESPEARE

One of the difficulties of deciding whether or not there exist laws of organization of a thermodynamic or related sort is bound up with a long-standing problem regarding the nature of time. Any organizational principle must, to be useful, tell us something about the development of complexity with time, but some would argue that in practice time might be nothing more than the ongoing development of certain types of organization. Whereas most physicists regard the second law of thermodynamics as a reflection of the improbability of certain types of initial conditions, there are others who regard it as a far more fundamental idea that is prior to the laws of Nature themselves. Moreover, it is only in situations where entropy changes are manifest that the notion of time becomes truly meaningful. Ilya Prigogine and Isabelle Stengers write:

> Only when a system behaves in a sufficiently random way may the difference between past and future, and therefore irreversibility, enter into its description... The arrow of time is the manifestation of the fact that the future is not given, that, as the French poet Paul Valéry emphasized, 'time is a construction'.

Yet, even if this were true, there still appears to exist something of a puzzle in a variety of areas.

In general, the laws of Nature that we believe we have found possess the property of time-reversibility. That is, if the laws permit a particular causal sequence of events—a history—then they will allow the time-reversed history also. Despite the ubiquity of this state of affairs amongst the laws of Nature, there exists an unmistakable predilection for Nature to exhibit histories of one directed type, never the reverse. This is sometimes called the 'reversibility paradox'. There are a number of particular physical phenomena that exhibit a directionality or 'arrow of time'. Part of the puzzle is to determine whether or not their individual directionalities are in any way related.

All radiation fields obey laws which permit what are called 'advanced' and 'retarded' solutions. The retarded solutions describe the appearance of a wave after the inception of its source, that is, spontaneous emission. The 'advanced' solution, on the other hand, describes a wave travelling from the future which is absorbed at the source. In reality, we observe only retarded solutions of the mathematical laws of wave propagation. Likewise, close to thermodynamic equilibrium, entropy and complexity increase with the passage of time. There exist equally permissible histories in which they decrease, but they are not observed. Decaying physical states, like radioactive nuclei, diminish exponentially with increasing time. And, last but not least, we possess a psychological sense of the passage of time. Our memory is of that part of time we call the past. It is clearly distinguished from the future.

We would like to know whether all these different senses of time direction are linked to each other and even linked to the global arrow of time provided by the expansion of the Universe. The conclusion of Stephen Hawking's widely

purchased book *A Brief History of Time* is that the psychological and thermodynamic arrows are the same because the brain is at root a computer and
computation is irreversible. The idea of this argument is to grant (although
some might be unwilling to do this) that the brain is just a computer which
carries out logical operations, and then argue that computation is irreversible
for thermodynamic reasons. Hence, mental processing possesses an arrow of
time endowed by thermodynamics. This argument is not convincing, because
computer scientists have shown that abstract computation is not logically
irreversible. While the operation of ordinary addition may be irreversible
(there is one way to add 3 + 3 to get 6, but 6 can be obtained from the addition
of 3 + 3, 4 + 2, 5 + 1 or 6 + 0) and the conventional computer's 'AND/OR' logic
gate clearly has one input and two possible outputs, it is none the less possible
to construct logic gates that are their own inverses. Computation using such
'Fredkin' gates are logically reversible and in ideal circumstances are not made
unidirectional by the second law of thermodynamics. This does not prove that
the thermodynamic arrows are not identical, only that this particular attempt
to prove them so fails.

FAR FROM EQUILIBRIUM

Here on the level sand
Between the sea and land,
What shall I build or write
Against the fall of night?
Tell me of runes to grave
That hold the bursting wave,
Or bastions to design
For longer date than mine.
—A. E. HOUSMAN

Dorothy Sayers' famous 'Peter Wimsey' story *Have His Carcase** was first
published in 1932 and was conceived during the period when the second law
of thermodynamics was rather fashionable amongst the chattering and writing
classes. Following the discovery of the body of a gigolo on an isolated English
beach, Wimsey hears the evidence of a series of witnesses and suspects. After

---

* For the benefit of non-English readers, the words of this title are Cockney rhyming slang for
'Habeus Corpus' an ancient English legal remedy to guard against wrongful arrest and prolonged
detention when adequate evidence is unavailable. 'You must produce the body . . . ,' as the original
Act of parliament states.

hearing that of Miss Olga Kohn, she thinks him a little sceptical of her, and asks:

> 'But you do believe me, don't you?'
>
> 'We believe in you, Miss Kohn,' said Wimsey solemnly, 'as devoutly as in the second law of thermo-dynamics.'
>
> 'What are you getting at?' said Mr. Simms suspiciously.
>
> 'The second law of thermo-dynamics,' explained Wimsey, helpfully, 'which holds the universe in its path, and without which time would run backwards like a cinema film round the wrong way.'
>
> 'No, would it?' exclaimed Miss Kohn, rather pleased.
>
> 'Altars may reel,' said Wimsey, 'Mr. Thomas may abandon his dress-suit and Mr. Snowden renounce Free Trade, but the second law of thermo-dynamics will endure while memory holds her seat in this distracted globe, by which Hamlet meant his head but which I, with a wider intellectual range, apply to the planet which we have the rapture of inhabiting. Inspector Umpelty appears shocked, but I assure you I know no more impressive way of affirming my entire belief in your absolute integrity.' He grinned. 'What I like about your evidence Miss Kohn, is that it adds the final touch of utter and impenetrable obscurity to the problem which the inspector and I have undertaken to solve. It reduces it to the complete quintessence of incomprehensible nonsense. Therefore, by the second law of thermo-dynamics, which lays down that we are hourly and momently progressing to a state of more and more randomness, we receive positive assurance that we are moving happily and securely in the right direction.'

In this account, we see a number of interesting perceptions of the second law. It is perceived as a true law which 'holds the universe in its path', rather than the consequence of a particular choice of initial conditions as we discussed in Chapter 3. More interesting is the assumption that time would run backwards if the law were reversed. The writer assumes the notion that the increase of entropy is such an overwhelming requirement that, were it to decrease with time, this could only mean that time had reversed its arrow. The other notion that permeates the dialogue is the belief that the second law requires everything, willy-nilly, to proceed towards a state of greater disorder. Hence, the increasingly confused and disordered state of the evidence available strikes a resonant chord in Wimsey's mind. But one wonders what he thought when the confusions were all finally ironed out and an orderly conclusion was drawn from the mass of conflicting stories.

The thermodynamic sense of order decrease that is enshrined in the second law is at first sight in conflict with many of the complicated things that we see going on around us. We see complexity and order increasing with time in many situations: when we tidy up our office, when we build a radio set out of a collection of pieces of wire and crystal, whenever a car company delivers a new car off the end of its production line, the evolution of complex life-forms from the simpler ones that the biologists tell us were our precursors. All these processes witness towards the possibility of passing from a state of relative disorder into one of considerable order.

In many of these cases, we must be careful to pay attention to all the order and disorder that is present in the problem. Thus the process of tidying the office requires physical effort on somebody's part. This causes ordered biochemical energy stored in starches and sugars to be degraded into heat. If one counts this into the entropy budget, then the decrease in entropy or disorder associated with the tidied desk is more than compensated for by the other increases.

However, there is an added subtlety when a system is a long way from being in a state of thermal equilibrium. In this situation, it will be sustained by some connection between an outside environment and its own internal organization. Far from equilibrium, unusual things can happen in the sense that our intuition about what is likely or 'probable' is largely conditioned by the so-called Gaussian law of large numbers derived from our experiences of what occurs very close to equilibrium. The study of systems far from equilibrium is still in its infancy. We have developed little intuition as to what is, and what is not, probable in complex natural phenomena given long periods of time over which events of very low probability can make their presence felt. A Theory of Everything alone cannot tell us what types of organized complexity exist in Nature. Such states are strongly conditioned by their detailed make-up and their actual histories. They may be governed by undiscovered general rules of evolution, distinct from the laws of Nature, which dictate the development of all forms of complexity. A Theory of Everything will make little or no impact upon such problems as the origin of life and consciousness. They sit on a different shelf in the storehouse of wonders.

As scientists have become more concerned with understanding the development of organized complexity, they have begun to use the term 'emergence' to describe the situation where significant levels of order arise from simple building blocks in a way that was not predictable at the level of the building blocks. Two hydrogen atoms and one oxygen atom make a molecule of water but that description is inadequate to foresee the existence of waterfalls and glaciers. Nature seems to create a staircase of increasing complexity so that each significant upward step is not fully reducible to the steps below. All the most interesting complex structures that we see around us are like that. The physiological make-up of a single human being is a matter for the biochemist and then the physiologist. Then, the neurophysiologist and the psychologist are needed. But, as soon as there is more than one person we need the sociologist, the economist, and the politician. And next, we need the theologian, the artist, and the musician. At each stage, a higher tier of complexity results that is not contained in the lower levels of behaviour. When we get to the higher levels of emergent complexity, we encounter an interesting multiplicity of descriptions that are different but complete within their own terms of reference. The display on your computer screen admits of a complete description at the level of electronics that makes no mention of the word-processing software that creates the document that is on display, and a description of that software would be regarded as complete even if it made no mention of the message that was actually on the screen. The description of its content and meaning are complementary to its full description at the level of atomic physics or electronic circuitry. No one would confuse one with the other in this example, but when it comes to talking about the 'meaning' of the Universe, and whether our descriptions of its structure and evolution are compatible with other attempts to understand the significance of human ethics and religious beliefs, these different categories are often forgotten, and understanding at one level is imposed upon another one.

THE SANDS OF TIME

Not all who wander are lost.
—J.R. TOLKIEN

Is it still possible to discover deep truths about the world by observing mundane objects? Or do fundamental discoveries always require millions of pounds, dollars, or euros and armies of people, along with huge particle colliders, batteries of computers, gigantic telescopes, or space satellites in orbit? On most frontiers, fundamental science has become big science. But there are some beautiful exceptions. One of the most impressive examples grew from careful thought about an observation that we have probably all made at one time or another. It has become a paradigm for the development of forms of complexity that appear to organize themselves out of disorder.

Create a pile of sand by letting it pour down under gravity on to a flat surface, like a small table top. The pile steadily builds up, gradually getting

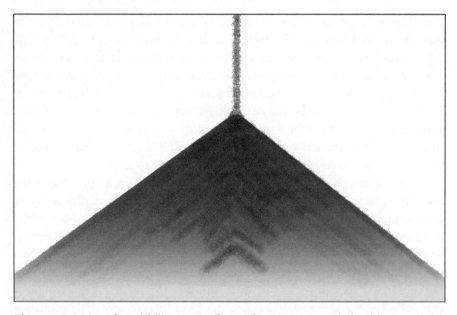

Figure 7.5 Grains of sand falling onto a flat surface create a sandpile with a particular slope. The critical slope is maintained by avalanches of all sizes.

steeper. Avalanches of sand continually occur as the sand pours down from above. At first, the infalling grains have effects only very close to where they fall but, as the pile steepens, the avalanches become more extensive in their effects. Eventually, something odd happens: the pile gets no steeper. A critical slope is achieved and adding more sand just produces a cascade of avalanches that maintain the same slope. If the pile is sitting on a table top then eventually the sand will start to flow over the edge of the table at the same rate that it falls in from above. The sandpile will present the same shape even through it is made up of different sand grains at different times, rather like a steadily flowing river (Figure 7.5).

What has happened is remarkable. Each of the infalling sand grains is following a chaotically sensitive trajectory in the sense that a small deflection on the way down from other grains results in a big change to its subsequent history—maybe falling down the other side of the pile. Yet, the net result of all these chaotically falling grains is a highly organized pile with a particular slope that depends only on the smoothness of the sand. Strangest of all, the pile maintains its organized slope by instabilities—avalanches—which occur on all dimensions from the size of a single grain right up to the length of the pile's slope.

This process was given the name 'self-organizing criticality' by Per Bak, Chao Tang, and Kurt Wiesenfeld, who first recognized its significance in 1987.

The adjective 'self-organizing' captures the way in which the chaotic sandy input seems to arrange itself into an orderly pile. The attribute of 'criticality' reflects the precarious state of the pile at any time. It is always about to experience an avalanche somewhere or other. The sequence of events that maintains its state of order is a slow local build-up of sand somewhere on the slope, then a sudden avalanche, followed by another slow build-up, sudden avalanche, and so on. When the local build-up of sand creates a little hill steeper than the critical slope a collapse occurs. Overall stability is maintained by local instability.

What is unexpected about this situation is that the pile continually evolves towards a precariously unstable state whereas most systems, like a ball rolling around inside a bowl, seek out the only stable resting place. The sandpile is increasingly susceptible to disturbances of all sizes as it nears its critical state, and always exists as a transient orderly state. If it forms on a table top, then sand arrives at the same rate that it falls off the edge of the table; the structure of the pile as a whole persists but is composed of different grains of sand. All that is needed for this type of critical structure to arise is for the frequency of an avalanche to depend only on a mathematical power of the size of the avalanche—that power will be negative as large avalanches are rarer than small ones. This means that there is no preferred size of avalanche. This wouldn't be true if the sand was sticky and tended to form balls of a particular size that just rolled down the side of the pile. Closer examination of the details of the fall of sand has revealed that avalanches of asymmetrically shaped grains, like rice, produce the critical scale-independent behaviour even more accurately because the rice grains always tumble rather than slide.

Originally, its discoverers hoped that the way in which the sandpile organized itself might be a paradigm for the development of most types of organized complexity. This was too optimistic. But it turns out to provide clues as to how many types of complexity organize themselves. The avalanches of sand can represent extinctions of species in an ecological balance, jams on a motorway traffic flow, the bankruptcies of businesses in an economic system, earthquakes or volcanic eruptions in a model of the pressure equilibrium of the Earth's crust, and even the formation of oxbow lakes by a meandering river. Bends in the river make the flow faster there, which erodes the bank, leading to an oxbow lake forming. After the lake forms, the river is left a little straighter. This process of gradual build up of curvature followed by sudden oxbow formation and straightening is how a river on a flat plain 'organizes' its meandering shape.

It seems rather spooky that all these completely different problems should behave like a tumbling pile of sand. A picture of Bak's (Figure 7.6), showing

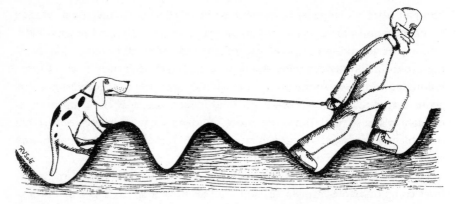

Figure 7.6 A realistic system with many possible local equilibrium states and a force which acts to move between them by slow hill climbing followed by sudden jumps.

a dog being taken for a bumpy walk, reveals the connection. If we have a situation where a force is acting (for the sandpile it is gravity, for the dog it is the elasticity of its leash) and there are many possible equilibrium states (valleys for the dog, stable local hills for the sand) then we can see what happens as the leash is pulled. The dog moves slowly uphill and then is pulled swiftly across the peak to the next valley, begins slowly climbing again, then jumps across. This staccato movement of slow build-up and sudden jump, time and again, is what characterizes the sandpile with its gradual build-up of sand followed by an avalanche. We can see from the picture that it will be the general pattern of behaviour in any system with very simple ingredients.

The nice feature of these insights is that they show that it is still possible to make important discoveries by observing the simplest everyday occurrences and asking the right questions.

THE WAY OF THE WORLD

I am sensible that this Tractate may likely incur the Censure of a superfluous piece, and myself the Blame of giving the Reader unnecessary Trouble, there having been so much so well written of this subject by the most learned men of our Time.
—JOHN RAY

The great unanswered question is whether there exists some undiscovered organizing principle which complements the known laws of Nature and dictates the overall evolution of the Universe. To be a true addition to what we know of Nature's laws, this principle would need to differ from any laws of gravitation and particle physics that might emerge in final form from some Theory of Everything. It would not be specific to universes but would govern the evolution of any complex system. True, its general notions ought to be tailored in some way to the notions which characterize the specific things that go on in an evolving Universe—the clustering of matter into stars and galaxies, the conversion of matter into radiation—but it would also need to govern the invisible ways in which the gravitational field of the universe can change. Any such discovery would be profoundly interesting because our Universe appears to be far more orderly than we have any right to expect. It has a tiny entropy level compared with the largest value that we could conceive of it possessing if we were to reorganize the observed matter into other configurations. This implies that the entropy level at the beginning of the expansion of the Universe must have been staggeringly small, which implies that the initial conditions were very special indeed. But this may be too simple a conclusion to draw. We have seen from our discussions of 'inflation' in the early universe that the part of the entire Universe which we now observe reflects the starting conditions of only a minute region of the whole universe of space. We cannot therefore draw any conclusions about the entropy of the whole Universe. Indeed, such a concept might not exist if the Universe is infinite in spatial extent. And the inflationary universe picture would lead us to believe that beyond our visible horizon it seems rather likely that things are rather disordered. From the thermodynamic point of view, we may therefore be a fluctuation.

Another curiosity about the entropy of the Universe relates to the traditional picture of the 'heat death' which would have us approaching closer and closer to a state of uniform temperature in the far distant future, after which nothing could happen. In fact, the situation is rather more complicated. It appears that even though the total entropy in the observed portion of the Universe is increasing and processes can be foreseen which will ensure that this increase continues unabated in the future, it is actually lagging farther and farther behind the theoretical maximum entropy level that it could in principle possess.

Elsewhere, Frank Tipler and the present author have examined the possible future histories that might ensue for the large-scale structure of the Universe in the light of known principles of physics. We were interested in discovering whether it is possible for some form of life to exist at all future times. In order to say anything meaningful about such a question, we have to cut it down to size in various ways. We do not know all the attributes of living things, so we shall focus upon those bare minima which would be necessary for intelligence to operate. In practice, this means that information processing must be able to occur and this requires some form of thermodynamic disequilibrium if it is to be possible. We are then able to show that there is no known obstacle to information processors of an appropriate type continuing to process information at all future times, or, more simply, that they can process an infinite amount of information in the unbounded future. This does not of course mean that they will, or even should; nor that such devices need possess any other properties that would identify them as living. The aim is to show that there is no obstacle to such information processing to the future; in particular, that it is not inevitably extinguished by the widely advertised 'heat death' of the Universe. This is the essential content of the so-called *Final Anthropic Principle*, or *Final Anthropic Conjecture* as it might be more appropriately termed. It is not a philosophical speculation, but a property that our particular Universe either does or does not possess. One might conjecture that if some grand organizing principle is discovered which governs the overall development of organized complexity even in whole universes then the answer to this Final Anthropic Conjecture will form part of it. Some measure of information processing capability and the algorithmic complexity and depth of information that can be produced might provide us with a candidate for our sought-after quantity. Indeed, these concepts have many attractive features to commend themselves. The notion of randomness will not be a fixed one in an expanding universe. As the available information grows and the computational complexity of natural information processors evolves, so the definition of what must be called random will evolve also.

If we were to regard the Universe as a vast computer, a processor of information, a generator of entropy, then we can readily envisage the laws of Nature as some form of software which runs upon the particular forms of matter that form the world of strings and elementary particles. A true unification of these two entities in the manner that we explored in earlier chapters would amount to a program that was very hardware specific. Such programs are easy to envisage. If we think of our own mental circuitry in this way, then it is clear that many of the brain's sub-programs are hardware specific: they move arms and legs and perform other specific motor functions. Initial conditions are akin to the initial input on which the program is going to act. If initial conditions must have special forms which are inextricably bound up with the laws and particles of matter, then this would require admissible universal programs to have only certain starting configurations in order to run. But we still seem to encounter something of an impasse, a 'dangerous loop'. It seems that our Theory of Everything must give rise to the concept and capabilities of such an abstract computer as well as be described by it.

The heat death of the Universe was a pessimistic outcome of the Victorians' thermodynamic contemplation of the Universe as a great machine, running down into a state of growing disorder and uninteresting equilibrium. The expansion of the Universe seems to offer little hope of escape. The closed universe that heads back towards a Big Crunch of ever-increasing density in the future is doomed to a finite future unless its inhabitants can find a way to process energy at ever faster rates as they head towards the crunch. If they succeed they can live for an infinite amount of their subjective time but have little scope for action. A universe that expands forever, as ours appears to be, has a number of possible fates. The simplest sort, which decelerates forever, has a way of avoiding the simple heat death. Even though its entropy must always increase, it turns out that it gets farther and farther away from the state of maximum entropy that is allowed in the Universe at any given time. The maximum allowed entropy increases faster than the actual entropy. Entropy is always increasing but we are always getting farther and farther away from equilibrium. However, our Universe does not seem to be on that sort of track. The discovery that the expansion of the Universe began to accelerate several billion years ago changes things in a major way. The acceleration means that there is actually an absolute maximum entropy for the Universe and our steadily increasing entropy can eventually reach it. When it does so, then we will have run into the heat death of the Universe. Our only hope is to stave it off locally by inhabiting local over densities of matter which do not get swept up to participate in the accelerating expansion of the Universe. But, while that may be possible for a time, eternity is a long time and eventually all these over densities will be ironed out and the Universe will be left featureless and lifeless forever, it seems.

CHAPTER 8

Selection effects

Don't bite my finger—look where it's pointing.
—WARREN S. MCCULLOCH

UBIQUITOUS BIAS

He who knows not and knows not that he knows not is a fool.
  Shun him.
He who knows not and knows that he knows not is a child.
  Teach him.
He who knows and knows not that he knows is asleep.
  Wake him.
He who knows and knows that he knows is a wise man.
  Follow him.
—ARAB PROVERB

No science can be founded upon observation alone. We would know neither what we were observing nor how our observations are biased by a propensity to gather some types of evidence more readily than others. As any good cross-examiner knows, certain types of evidence are more readily obtained than others. Consequently, the mark of a good experimentalist is not just a practical dexterity but the ability to understand and forsee as completely as possible any biases that are innate to the types of experiment and observation that he employs.

Such biases play a crucial role in our attempts to understand the Universe in its totality. Any Theory of Everything that ignores the influence of bias will fail to make accurate contact between its predictions and what is actually seen in the Universe. A complete understanding of our observations of the Universe requires us to take into account those errors which are introduced by the act of observership.

Scientists are familiar with two types of experimental 'error'; neither has anything necessarily to do with the everyday sense of the word 'error'. The first is the limiting accuracy with which a measurement can be made. This form of error always exists at some level and the aim of the scientist is to minimize it. The second variety of error—'systematic error'—is more subtle and not necessarily avoidable. Every scientific procedure will contain some tendency to skew the results in one direction or another. In laboratory experiments, there is the possibility of repeating experiments with certain of the ambient conditions changed to investigate whether the results depend upon some of those conditions. Scientists like important discoveries to be confirmed by at least two independent experiments for the simple reason that each will have different systematic biases because their instruments will never be completely identical. However, in astronomy we are less fortunate. We can observe the Universe, but we cannot alter its configuration so as to carry out controlled sequences of experiments upon it. We cannot carry out all possible experiments or record all possible data. We are faced with a confinitive rather than an infinitive system and therefore we must be especially aware of all the possible biases that render certain observations inevitable. Thus, if we were to commission a survey of all the visible galaxies with a view to determining their relative brightness, we would have to deal with the in-built bias towards finding the brighter galaxies more easily than the fainter ones.

In cosmology, this type of selection bias is all-pervading, and a recognition of the fact is enshrined in what has become known as the Weak Anthropic Principle. This is most usefully viewed as the recognition that our own existence requires certain necessary conditions to be met regarding the past and present structure of the visible universe. Our observations must not be viewed as having been taken from some unconstrained ensemble of possibilities but from some subset conditioned by the necessary conditions for carbon-based observers like ourselves to have evolved before the stars die. Cosmologists view the Weak Anthropic Principle as a qualification of the famous stricture of Copernicus that man does not occupy a special position in the Universe. For, although we are right to disregard the prejudice that our position in the Universe is special in *every* way, we should not conclude from this that our position cannot be special in *any* way. We could not exist within a star; we could not exist when the Universe was less than a million years old and temperatures were high enough to dissociate any atom or molecule. If the Universe did happen to possess a centre (there is no evidence that it does) and conditions were only conducive to the evolution and continued existence of life near that centre, then we should not be surprised to find ourselves living there. One of the most important features of the Weak Anthropic Principle is that its disregard will lead to erroneous conclusions being drawn about the structure of the Universe. The most notable example is that of Dirac who was misled into proposing a very radical change to the law of gravitation in order to explain a numerical coincidence between constants of Nature and the age of the Universe because it was not recognized that this coincidence was a necessary condition for the existence of observers.

The Universe, it was once assumed, existed within the framework of some vast unchanging background of space upon which all the observed motions of the heavenly bodies are played out. We have discovered that there is no such static cosmic stage. Everything that is—the entire visible universe of stars and galaxies—is in a state of perpetual motion. The Universe is expanding: its clusters of galaxies are flying away from each other at a speed that increases in proportion to their separations. This cosmic recession is revealed to us by the systematic redshifting of the light from distant sources.

If we retrace the course of this expansion backward in time, we can visualize an apparent beginning to the current state of expansion, about fifteen billion years ago, when all separations extrapolate back to zero. Current cosmological research focuses upon events during the first fraction of a second after the apparent beginning. In these moments, the Universe resembled a cosmic experiment in high-energy physics, the fall-out from which enables us partially to reconstruct its structure.

The problem of fitting human life into the impersonal tapestry of cosmic space and time has been pondered by mystics, philosophers, theologians, and scientists of all ages. Their views straddle the entire range of options. At one extreme is painted the depressing materialistic picture of human life as a local accident, totally disconnected and irrelevant to the inexorable march of the Universe from the 'Big Bang' into a future 'Big Crunch' of devastating heat, or the eternal oblivion of the 'heat death'. At the other is preached the traditional teleological view that the Universe has some deep meaning, and part of that meaning is ourselves. On this optimistic view, we might not be surprised to find our local environment tailor-made for our needs. This latter view remained that of many biologists until, in the middle of the nineteenth century, Charles Darwin and Alfred Russel Wallace made their crucial observations and deductions about the evolutionary adaptation of organisms to their environment. Since that time, biologists have rejected any notion that evolution is goal-directed in any way. If the environment were to change in some unusual fashion, so as to render intelligence a liability, then we would find ourselves following in the distinguished footsteps of the dodo and the dinosaurs.

Cosmology does not have anything interesting to say about the detailed functioning and evolution of terrestrial life, but it does have some surprising things to say about the necessary prerequisites for it. Let us take a simple but striking example. The visible universe is about fifteen billion light years across. It contains at least one hundred billion galaxies, each of which contains about one hundred billion stars like the Sun. Why is the Universe so big?

Living systems on Earth are based upon the subtle chemical properties of carbon and their interplay with hydrogen, nitrogen, phosphorus, and oxygen. These biological elements, and all much-vaunted alternatives like silicon, do not emerge as fossils from the inferno of the Big Bang. They are the results of nuclear reactions in the interiors of the stars. There, primordial hydrogen and helium nuclei are burnt into heavier elements by the process of nuclear fusion. When these stars reach the ends of their lives, they explode and disperse these heavier biological elements into space where they become incorporated into molecules, planets, and eventually people. Almost all the carbon atoms in our bodies share this dramatic astral history.

This process whereby Nature produces the biological building blocks of life from the inert relics of the Big Bang is long and slow by terrestrial standards. It takes more than ten billion years. This vast period of stellar alchemy is necessary to provide the necessary precursors to life. Since the Universe is expanding, we now discern why it is necessary for it to be at least ten billion light years in size. A universe simply as big as our Galaxy indeed has room for a hundred billion stars, but it would be little more than a month old. There is a niche in the history of the Universe when life could and did evolve spontaneously. That niche is bounded on one side by the requirement that the Big Bang cool off sufficiently to allow stars, atoms, and biomolecules to exist, and on the other by the fact that all the stars will have burned out after a hundred billion years (see Figure 8.1(a)).

The simple lesson to be drawn from this example is that the large-scale structure of the Universe is unexpectedly bound up with those conditions necessary for the existence of living observers within it. When cosmologists are confronted with some extraordinary property of the Universe, they must temper their surprise by considering who would be here to be surprised if the Universe were significantly different. This type of 'Weak Anthropic' consideration is not a falsifiable conjecture or a theory. It is an example of a methodological principle which, if ignored, will lead one to draw incorrect conclusions from the data at hand.

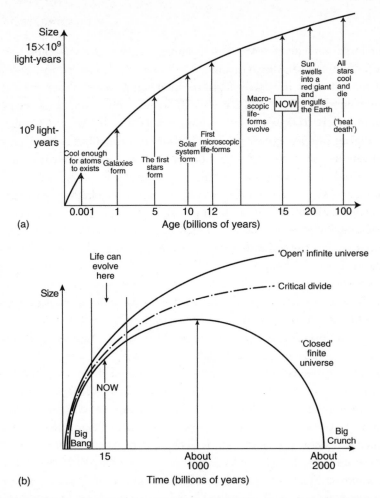

Figure 8.1 (a) The characteristic epochs of cosmic history in an expanding universe which never accelerates. The expansion means that the ambient conditions of density and temperature change continuously with time. Only after sufficient periods of time have passed are conditions cool enough for the formation of atoms, then molecules, then stars, planets, and life. To the future, we foresee an epoch when the stars will all have exhausted their nuclear fuel. If carbon-based life-forms did not evolve within the niche indicated, then they will never evolve. (b) Our Universe expands very close to the critical divide which separates those universes which will expand forever from those which will eventually collapse back towards a Big Crunch of ever-increasing density. Only those universes (like our own) which start expanding very close to the critical divide will give rise to biochemical complexity and observers at some stage in their histories: those that start expanding too slowly will collapse and return to a Big Crunch before the temperature falls sufficiently for stars, or even atoms, to form; those that start expanding too rapidly prevent galaxies and stars from ever forming because the force of gravity never halts the expansion in a local region, and the absence of stars precludes the production of the heavy elements that are necessary for the spontaneous evolution of life. If acceleration occurs (as shown in Figure 5.2) then all galaxy formation will cease once the accelerated expansion commences.

The impact of ignoring this principle will depend upon the basic structure of the Universe. If there exists some intrinsically random element in the make-up of the Universe, then the role of bias becomes crucial to our programme of understanding the physical world. If the Universe has one necessary and unique possible structure because there is only one possible logically consistent universe, then our Weak Anthropic selection effect allows us to conclude little more than our good fortune that 'the' Universe happens to allow life to evolve. However, we appreciate from our discussion of the role of symmetry-breaking in Nature that the Universe does not seem to be like this. There exist aspects of the Universe which could have been otherwise, and indeed may actually be otherwise in different parts of the cosmos. Moreover, we saw that many constants of Nature owe their values to quasi-random processes occurring in the earliest stages of the Universe. In such circumstances, we would make a grave error if we were to expect that the predictions of the most likely universe to emerge from the Theory of Everything must correspond necessarily to the one we see.

Earlier, we pointed out that, in controlled terrestrial experiments, we can repeat our observations with various conditions altered. Hence, it is often straightforward to elucidate which phenomena distinguish features intrinsic to the laws of Nature from those which are merely the consequence of some symmetry having broken one way rather than another. When we enter the astronomical realm things are not so clear-cut. We do not know, for example, whether the sizes of galaxies and galaxy clusters are fundamental consequences of physical laws, of special initial conditions, or of some symmetry-breaking process having fallen out in one particular way. The only substitute we have for unhindered experiment is to compile catalogues of all the observable properties of collections of similar objects and then search for *correlations* between different quantities. Thus we can uncover trends: see whether all big galaxies are bright or whether all magnetic stars rotate slowly, and so forth.

Until only a few years ago, the influence of random symmetry-breaking upon the observed structure of the astronomical universe of galaxies and clusters was regarded as somewhat speculative with no basis in the favoured picture of the evolution of the very early universe. This has now changed. The gradual maturing of the 'inflationary universe' hypothesis, which we introduced in earlier chapters, makes the idea of a quasi-random aspect to the early universe appear a very natural one. For, if the Universe begins expanding from a state in which conditions vary from place to place, say in a random fashion, then different microscopic regions will inflate by different amounts; that is, they will each undergo periods of inflation of different length. Only those regions which inflate for long enough and subsequently give rise to regions large enough for atoms, stars, and hence life, to evolve will be sites for subsequent cosmological speculation.

When we come to compare the predictions of this theory with observation and to understand the structure of the observed Universe in terms of this chaotic inflationary universe theory, we need to take into account the bias that is present in our observations of the Universe. Observations can only have been made in particular types of universe. We would not be justified in excluding this theory from consideration on the grounds that the majority of the inflated regions are tiny. We would have to be living in one of the large ones regardless of how low its *a priori* probability might be. Moreover, if the Universe is spatially infinite, then our observations of a particular habitable backwater make the extrapolation to grandiose conclusions about the nature of the Universe as a whole precariously dependent upon untestable assumptions about the nature of the Universe beyond our visible horizon (see Figure 8.2).

There is a further refinement of this chaotic inflationary picture of the early universe, suggested by the Soviet physicist Andrei Linde, in which the process of inflation is self-perpetuating. Each microscopic region that inflates tends naturally to recreate the conditions for its own microscopic sub-regions subsequently to inflate, and the process need never end. By the same token the region which you imagined as being the starting point for this sequence could form part of a past infinite sequence. Only in those members of the infinite sequence where the necessary conditions for the evolution of observers are met will cosmological deductions be drawn. The scenario of eternal inflation is illustrated in Figure 8.3.

The influence of the Weak Anthropic Principle has grown as cosmologists have probed closer to the initial state in their attempts to reconstruct the past history of the Universe. The closer one approaches to the apparent beginning, so the effects of symmetry-breaking and quantum randomness begin to proliferate and generate the intrinsically random elements whose legacy creates the subtleties of interpretation that we have highlighted.

These problems introduce a difficult challenge into our attempts to understand the nature of the Theory of Everything and its cosmological consequences. When a theory has a range of possible outcomes, either because of symmetry-breaking or its quantum character, then comparing it with observations is subtle. We need to recognize that only some outcomes will allow intelligent beings to exist, and we must be living in one of those outcomes, no matter how improbable it might be *a priori*. Otherwise, we risk mistakenly ruling out the theory because it predicts that a universe like the one that we

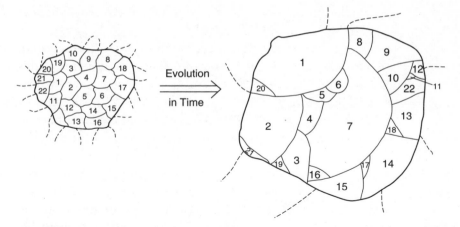

Figure 8.2 The evolution of a chaotic inflationary universe. Every microscopic causally connected sub-region of the universe of size 10^{-25} centimetres inflates by a different amount when the expansion has continued for about 10^{-35} seconds. Each of these (numbered) sub-regions grows into a corresponding large region like our visible universe today. Only in those regions which inflate sufficiently to remain expanding close to the critical divide separating indefinite future expansion from eventual collapse (see Figure 8.1) can give rise to intelligent observers. Hence, life can only arise in the largest inflated regions like Region 7. If the universe is infinite in size then there will exist an infinite number of these regions, and if their initial properties randomly exhaust all possibilities then there will arise an infinite number of regions in which conditions are suitable for living observers to exist. Hence, if there is any finite probability that life can evolve (and clearly there is because we, at least, are here), then it must have done so at an infinite number of these sites elsewhere in a universe of infinite size. Notice how this alters our picture of the nature of the Universe. If we reside in Region 7, then beyond our visible horizon conditions would be expected to be very different. Observations of our observable portion of the whole Universe may be singularly unrepresentative of the whole.

see arises only with very low probability. In order to carry out this appraisal we need to know all the consequences of a Theory of Everything that impinge upon the existence of life. This is a tall order. In practice, only a few simple consequences are usually evident—no atoms can exist or no stars, say—but in the future much more complicated consequences will have to be evaluated in this way.

An interesting example of how this type of reasoning affects what a theory can tell you was provided a few years ago by applying it to a speculative proposal made by the American physicist Lee Smolin. Taking up a general suggestion of John A. Wheeler's, Smolin suggested that the 'constants' of Nature might be slightly shifted each time that matter collapsed to form a black hole

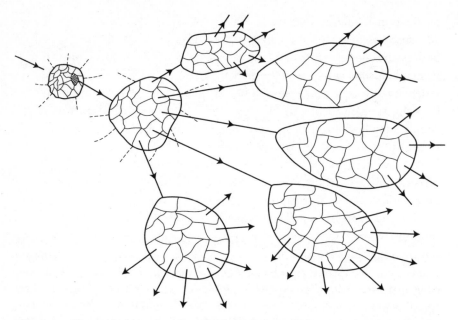

Figure 8.3 The evolution of an 'eternal' inflationary universe. Each sub-region that inflates can give rise to a large number of inflated regions which meet the conditions necessary for them to undergo further inflation themselves. This process can continue *ad infinitum* and, by the same token, may have been continuing since a past eternity also. Coupled with the scenario of Figure 8.2, we see that there can exist an infinite sequence of inflationary universes in time extending over infinite space. Only in some of those inflated outcomes, at particular times and in particular places, will the inflation proceed sufficiently to give enough time and the right conditions for life to be able to evolve, and only in some of those favoured regions will things fall out in such a manner that life actually does evolve successfully.

because the infalling material 'bounced' back to create a new universe from the singularity at the black hole centre. As a result, in the long run the most likely universe to be in is one in which the constants shift so as to maximize the formation of black holes. Thus, it was claimed that we could test this theory by carrying out some thought experiments. Any small change in the actual values of the constants of Nature should always *reduce* the mass of black holes in the universe.* However, we can see that (regardless of whether this is actually true for all changes we can imagine) it is not really what the theory predicts. We can only find ourselves in one of the universes that maximizes black-hole

---

* This assumes, of course, that there is such a local maximum for the black-hole production. Some constants, like the one controlling the strength of gravity appear to possess no such local maximum and black-hole production could be arbitrarily large for arbitrarily small values of that constant.

production if such a universe allows life to exist. Although this theory was advertised as being an alternative to the Anthropic Principle with regard to understanding the values of the constants of Nature, it requires the Anthropic Principle just as surely as any other theory, and it cannot make predictions independently of it. Sometimes the Anthropic Principle is wrongly believed to be a new type of 'theory' of the Universe. It is really nothing of the sort. It is simply a methodological principle which, if you ignore it, you run the risk of ruling out the correct or best fit cosmological theory.

We have seen that a complete understanding of our observations of the physical Universe requires an understanding of those elements which bias our observations and interpretations of data. If the Universe possesses intrinsically random elements in its make up, inherited from its quantum origins or from random symmetry-breakings during its early evolution, then we must take our own existence into account when evaluating the correspondence between reality and the cosmological predictions of any Theory of Everything. More-over, if these random cosmological elements lead to a universe which differs significantly from place to place over the very large distances, then our local observations of a possibly infinite Universe will inevitably leave our knowledge of its global structure seriously incomplete.

CHAPTER 9

Is 'pi' really in the sky?

Behold the heaven, the earth, the sea; all that is bright in them or above them; all that creep or fly or swim; all have forms because all have number. Take away number and they will be nothing...Ask what delights you in dancing and number will reply: 'Lo, here I am.' Examine the beauty of bodily form, and you will find that everything is in its place by number. Examine the beauty of bodily motion and you will find everything in its due time by number.

—ST AUGUSTINE

IN THE CENTRE OF IMMENSITIES

I would not go so far as to say that to construct a history of thought without a profound study of the mathematical ideas of successive epochs is like omitting Hamlet from the play which is named after him. That would be claiming too much. But it is certainly analogous to cutting out the part of Ophelia. This simile is singularly exact. For Ophelia is quite essential to the play, she is very charming—and a little mad. Let us grant that the pursuit of mathematics is a divine madness of the human spirit, a refuge from the goading urgency of contingent happenings.

—ALFRED NORTH WHITEHEAD

What is man that he is mindful of the Universe? Given the centuries of human history during which we have been ignorant of the vast oceans of outer space and the entire inner space of elementary particles, we recognize the twentieth century as a turning point in our appreciation of the breadth and depth of the structure of the Universe. Our quest for some ultimate explanation of the Universe's origin and structure witnesses to an unquestioned belief in our ability to understand the basic fabric of reality. But how strange this is.

Our minds are the products of the laws of Nature; yet they are in a position to reflect upon them. How fortuitous that our minds (or at least the minds of some) should be poised to fathom the depths of Nature's secrets. This fortuitous circumstance has two strands: one quantitative and one qualitative. The quantitative aspect is obvious: why should *we* be clever enough to fathom the Theory of Everything? We know of mathematical theorems which are undemonstrable in principle and others that would take our fastest computers the entire age of the Universe to decide. Why should the Theory of Everything be simpler than these? At root, these quantitative limitations are dictated by the size of the human brain, the capacity of our memories, or the capabilities of any other artificial brains that we might be able to fabricate. As yet, we do not know whether there exist fundamental limits upon the capacities of brains and computers (viewed as information-gathering and utilizing systems) which are imposed by the laws of Nature. Very likely there are. For, if we build a larger and larger artificial brain like a computer, then it grows in volume of circuitry faster than does the area of its surface and it is this area of surface that determines how effectively it can radiate away waste heat so as to avoid over-heating. To side-step this drawback of becoming large, one could take a leaf out of the natural world and develop the crenellated structure of a sponge, so displaying a far larger surface area than would a solid object of the same mass and volume. But this strategy greatly increases the length of circuitry that is necessary to keep the entire system coordinated, and thus slows down the rate at which it can send signals from one side of itself to the other.

A more interesting problem is the extent to which the brain is *qualitatively* adapted to understand the Universe. Why should its categories of thought and understanding be able to cope with the scope and nature of the real world? Why should the Theory of Everything be written in a 'language' that our minds can decode? Why has the process of natural selection so over-endowed us with mental faculties that we can understand the whole fabric of the Universe far beyond anything required for our past and present survival?

There is one qualitative aspect of reality that sticks out from all others in both profundity and mystery. It is the consistent success of mathematics as a description of the workings of reality and the ability of the human mind to discover and invent mathematical truths. And it is this mystery that we shall now endeavour to explore, because it draws us closer to the puzzle of why the Universe is intelligible at all.

THE NUMBER OF THE ROSE

God is more like gravitation than embarrassment.
— MARY HESSE

'A rose by any other name would smell as sweet', but not a rose by any other number. We have learnt that there is a profound difference between words and numbers. If you call a rose a thistle, then one is not seeking to rewrite any intrinsic property of those things we call roses: at worst a few horticultural catalogues would need revising, but the nature of things is not being tinkered with. But, if something has a numerical property, then to change it requires a deep and profound perturbation to the bedrock of reality. This impression is created by the assumption that mathematical properties of things are real and intrinsic to them. They are more than labels. We discover them; we do not merely invent them. Moreover, although we use language to describe the world, there does not seem to exist any natural correspondence between the rules of grammar and composition which dictate how the language is to be used. Yet, mathematics is a language that possesses a built-in logic which is unexpectedly attuned to the logic of reality.

Modern science is founded almost entirely upon mathematics. This preoccupation with the numerical as the route to understanding the physical seems to have begun with the Pythagoreans' conviction that the true meaning of Nature was to be found only in those harmonies that numbers display. Their view of creation was formed by a notion of the basic unit from which all other things could be constructed. Numbers had deep significance. Even numbers were seen as feminine and symbolized those things that belonged to the mother Earth; odd numbers were masculine and associated with the heavens. Individual numbers had meanings: four was justice, five marital union, and so on. We are heirs to this predilection towards numbers, but we have found it expedient to deviate from it in one crucial respect. Whereas the Pythagoreans were persuaded that numbers themselves were possessed of some especial significance, we have found it more fruitful to place significance upon the fact that there exist numerical relationships between things. Thus, we focus attention upon 'symmetries' and 'transformations', or 'mappings' and 'programs'. This approach matured most successfully in parallel with the mechanical world-view engendered by the work of Galileo, Newton, and their like-minded disciples. The mathematical description of Nature enabled human thinking about Nature to transcend cultural bias by identifying the irreducible minimum that characterizes the lawfulness of Nature. It created a universal language that aided efficient thought and deduction by building into its very fabric a number of simple logical requirements which will automatically be fulfilled whenever the language is used. In effect, it removes a number of logical operations from the realm of the conscious to that of the subconscious mind.

Seen in this light, mathematics may seem something of an art-form, and indeed in some universities it might be associated more closely with the arts and humanities than with the sciences. Yet it differs from the arts in many intriguing respects. Mathematics exhibits simultaneous discovery, whereas the arts do not; indeed, intuitively, we might feel that they cannot. Independent mathematicians working in different cultures, feeling different motivations, using different notations and methods, often produce the same final discoveries or 'theorems'. Such coincidences do not happen in literature or in music. The independent discovery of *Macbeth* or a Beethoven symphony would be inconceivable, because we associate so much of their essential nature with the mind of their creator. Their uniqueness is a reflection of the uniqueness of the individual. The fact that simultaneous discovery occurs in mathematics, as well as the sciences, points towards some objective element within their subject matter that is independent of the psyche of the investigator. We would confidently expect an intelligent machine to prove theorems that were similar, and in some cases identical, to those found by human mathematicians. Another interesting contrast between mathematics and the subjective humanities is in the working habits of their practitioners. It is common for mathematicians and mathematical scientists to work in collaboration. Many research papers in these fields have multiple authors. In some cases, these collaborations reflect the complementary nature of the collaborators' skills—one may be good at formulating interesting problems, whilst another might possess a greater talent for the technical implementation of their solution—but in many others no such straightforward demarcation exists. All the authors will contribute at all levels and distil their final results through a continual process of interaction or dialogue. On a personal note, I have written joint research papers with two individuals whom I have neither met nor even spoken to over the telephone. Such collaboration is rare in the humanities. There are famous collaborations, like that of Gilbert and Sullivan, but these are invariably the pooling of distinct skills. In the case of Gilbert and Sullivan, one created the music, the other the lyrics, for their operettas. How many novels or works of art can you name with multiple authorship? Again, one might suspect that this is indicative of the intrinsic subjectivity, and hence uniqueness, of artistic creation. The ease with which collaboration occurs in mathematical research and the essential similarity of the fruits of such collaboration to that of individual work points suggestively towards a powerful objective element behind the scenes that is discovered rather than invented.

PHILOSOPHIES OF MATHEMATICS

To abrogate the physical laws as in the scriptural miracles did not worry the religious philosophers as much as the abrogation of the mathematical laws. Thus, mathematics is accorded a distinguished position, and the possibility of its eternal truths being abrogated, even by an omnipotent God, is disturbing.
— PHILIP DAVIS AND REUBEN HERSH

Mathematics is a science of things thought of. Not everyone is persuaded that mathematics is merely discovered, nor even that the only alternative is for it to be a human mental creation. And, so, before we look at the unreasonable effectiveness of mathematics in accounting for the workings of the physical world, it is good to have before us the options that are on offer regarding the nature of this 'thing' that we call mathematics. When we learn or teach it, we never seem to ask, far less answer, this deceptively simple question. Yet it is not a new question and it is interesting to highlight some of the issues that coloured the discussion of it in the distant past when the wider climate of presupposition about the world drew its breath from very different sources than at present. A good flavour of some of the issues at stake can be gained from taking snapshots from three epochs when debate about the nature of mathematical knowledge was particularly intense. The first encompasses the argument between the Platonic and Aristotelian views in ancient Greece. The second is the voluminous commentary of Roger Bacon and his medieval contemporaries. And the last covers the joint developments of mathematics and physics at the end of the nineteenth century.

Plato argued that the material world of visible things was but a shadow of the true reality of eternal forms. He proceeds to explain the nether world of eternal blueprints most completely in the case of the elements of matter: earth, air, fire, and water. These he represents by geometrical solids: the earth by a cube, water by an icosahedron, air by an octahedron, and fire by a tetrahedron. His position is that ultimately the elements *are* just these solid geometrical shapes not simply that they possess geometrical shapes as one of their properties. The transmutation of elements one into the other is then explained by the merger and dissolution of triangles. This strictly

mathematical description characterizes Plato's discussion of many other physical problems. For him, mathematics is a pointer to the ultimate reality of
the world of forms that overshadows the visible world of sense data. The
better we can grasp it, the closer we can come to true knowledge. Thus, for
Plato, mathematics is more fundamental, truer, closer to the eternal forms of
which the visible world is an imperfect reflection, than the objects of physical
science. Because the world *is* mathematical at its deepest level, all visible phenomena will have mathematical aspects and be describable by mathematics
to a greater or lesser extent, determined by their closeness to their underlying
forms.

Aristotle's later view of the relationship between mathematics and Nature
could not have been more different. He wanted to rescue physical science from
the mathematical stranglehold that Plato had placed upon it. He believed
there to exist three completely autonomous realms of purely theoretical
knowledge—metaphysics, mathematics, and physics—each possessing its own
methods of explanation and accordant subject matter. But over-arching these
divisions there existed a more general principle of 'homogeneity'—that like
follows like—which must always be obeyed:

> It seems that perceptible things require perceptible principles, eternal things
> eternal principles, corruptible things corruptible principles; and, in general, every
> subject matter principles homogeneous with itself.

Plato's explanation of things clearly violates this principle in seeking mathematical explanations of physical things rather than physical explanations of
physical things. To understand Aristotle's view of the relationship between
mathematics and physics, we must appreciate that his threefold division of
theoretical knowledge into metaphysics, mathematics, and physics was hierarchical and quite different from Plato's treatment of the same three pillars
of knowledge. Physics deals with the ordinary everyday world of tangible
things devoid of any theoretical abstractions. It is the realm of the pragmatist.
Mathematics comes to deal with things only after one level of refinement
has been attained, by abstracting certain essential properties of things and
neglecting others. Finally, the neglect of all properties except that of pure being
is held to be the result of the further level of abstraction required to take our
study into the realm of metaphysics. Today, we might set up a comparable
hierarchy in terms of the outworkings of laws of Nature, the laws of Nature
themselves and then the meta-world in which we consider various possible or
actual alternative laws of Nature.

Aristotle draws a sharp dividing-line between the activities of the physicist and those of the mathematician. The mathematician limits his enquiry to the quantifiable aspects of the world and so dramatically restricts what is describable in mathematical terms. Physics, for Aristotle, was far wider in scope and encompassed the earthy reality of sensible things. Whereas Plato had maintained that mathematics was the true and deep reality of which the physical world was but a pale reflection, Aristotle claimed mathematics to be but a superficial representation of a piece of physical reality. Such is the contrast between idealism and realism in the ancient world.

In the Middle Ages, this conflict between the Platonic and Aristotelian views of the relationship between mathematics and the world began to re-emerge after the sleep of centuries. The question became intricately entwined with the labyrinthine syntheses of Aristotelian and Platonic ideas within early Christian theology. Influential thinkers like Augustine and Boethius implicitly supported the Platonic emphasis upon the primary character of mathematics. Both of them pointed to the fact that things were created in the beginning 'according to measure, number, and weight' or 'according to the pattern of numbers'. This they took to exhibit an intrinsic feature of the mind of God and thus mathematics took its place as an essential part of the medieval quadrivium without which the search for all knowledge was impaired. Yet Boethius later veered towards the Aristotelian viewpoint that some act of mental abstraction occurs *en route* from physics to mathematics which renders these two subjects qualitatively distinct.

The twelfth century saw a resurgence of scholarship and inquiry. There was interest in both the Platonic and Aristotelian perspectives on the relationship between mathematics and the physical world. The foremost commentators on the subject for the next century would be the English scholars Robert Grosseteste and Roger Bacon. Grosseteste argued that not all knowledge of the physical world relies upon mathematics, and on many occasions seems merely to echo the traditional Aristotelian position. But he went somewhat further, pointing out how some sciences are subordinate to others and, in his detailed studies of light, stressed that mathematics was essential for an explanation of what was seen 'since every natural action varies in strength and weakness according to variation of lines, angles, and figures'. Grosseteste influenced Roger Bacon's ideas about mathematics and Nature. Bacon wrote hundreds of pages on the subject and, indeed, no historical figure has ever appeared more preoccupied with the question than he. He believed that mathematical knowledge was innate to the human mind and mathematics was a unique form of thought known both by ourselves and by Nature. Its uniqueness is characterized by the fact that it allows complete certainty to be achieved and hence our knowledge of Nature can be secure only in so far as we found it upon mathematical principles:

> Only...mathematics remains certain and verified to the limits of certitude and verification. Therefore all other sciences must be known and certified through mathematics.

Moreover, Bacon was adept at using mathematics to prove the properties of the Universe. The most intriguing are the first 'topological' arguments about the nature of the Universe which he gives. He argues that the Universe must be spherical otherwise its rotation would create a vacuum. Furthermore, there can exist only one Universe because, were there another, it would, by the same token, need to be spherical and there would then be an anti-Aristotelian void between it and 'our' Universe. Bacon's position is midway between that of Plato and Aristotle and owes much to Grosseteste. He allowed mathematics a wider role in things without regarding it as the seat of all being. In practice, he made effective use of mathematics both in practical science and in the defence of his religious ideas.

Despite the legacies of Galileo and Newton's mathematical study of Nature, a sceptical philosophical tradition on the continent of Europe ensured that by the nineteenth century the rapidly blossoming field of mathematics was viewed as having a decreasing relevance for physical science. Mathematics underwent a dramatic expansion, but began to divide into the categories of so-called 'pure' and 'applied' mathematics. Influential physicists like Drude and Kirchhoff maintained that the task of science was to *describe* how the world was as simply and completely as possible. Science, they argued, can tell us nothing about reality: it is only 'a representation of the world of phenomena'. Indeed, Drude argued that there was a real danger in believing the world to be intrinsically mathematical since we could be led blindly into error by the rigid formalism of the pure mathematicians. Such views were not uncommon. Besides being voiced by operationalist philosophers they were also shared by physicists like Maxwell, Hertz, Boltzmann, and Helmholtz. With this background, there then arose in the early years of the twentieth century a debate as to the meaning and relevance of Leibniz's old notion of 'pre-established harmony' between the mathematical intuitions of the mind and the structure of the external world. This, in the more abstruse philosophical vocabulary of the time, is the question that we now raise as to the unreasonable effectiveness of mathematics in the description of the physical world.

Leibniz had wished to arrive at a convincing explanation for the harmonious relation between the capabilities and perceptions of our minds and the structure of the physical world of experience. This constituted a problem because he held these two realms, of mind and matter, to be totally separate. To resolve the tension, he propounded the idea that there exists a 'pre-established harmony' between the two realms.

There were many reactions to this question at that time. Some, like Fourier, had urged that mathematical knowledge should be obtained primarily from the study of Nature. The pre-established tripartite harmony between mind and mathematics and the physical world was supported by Hermite, who saw a metaphysical identity between the world of mathematics and physics that the mind shared in. Today, the notion of pre-established harmony seems little more than a disguised version of Platonism. Implicitly, it points to abstract mathematical notions that are the source both of our mathematical ideas and the mathematical aspects of the physical world. Both are reflections, albeit of differing intensity, of the mathematical blueprints that reside in the Platonic heaven. But, as powerful mathematicians like Minkowski and Hilbert found striking harmony between their pure mathematical results and the workings of the physical world, many found the claims for such a harmony hard to resist. Thus, in the early years of the twentieth century, we begin to see why Minkowski's application of complex numbers to the description of space and time was hailed by one physicist as 'one of the greatest revolutions in our accepted views'.

The puzzle that presented itself was to what extent the particular aspects of the real world necessary to identify its uniqueness—the precise values of constants of physics, the choice between one form of equation and another— are needed in addition to mathematics. Although a large part of a physical theory like Einstein's general theory of relativity appears to be mathematics and only mathematics, none the less, the coupling between matter and space-time geometry is not dictated by mathematics alone: it must incorporate the conservation of energy and momentum. Moreover, there is no known reason why the geometry of space and time should be described by the particular types of curved geometry defined by Riemann. There exist other more complicated varieties that could in principle have been employed by Nature. Only observation can at present tell us which has been used. Thus mathematics is unable to tell us which mathematics is chosen by Nature for employment in particular situations. This may of course merely be a transient manifestation of our relative ignorance of the bigger picture in which everything that is not excluded is demanded.

Let us give the last word on the notion of a pre-established harmony, which so exercised the early twentieth-century physicists, to Albert Einstein, who, as a young man, was a true believer in Leibniz's explanation of why Nature had to conform to abstract human thought:

Nobody who has really gone deeply into the matter will deny that in practice the world of phenomena uniquely determines the theoretical system, in spite of the fact that there is no logical bridge between phenomena and their theoretical principles; this is what Leibniz described so happily as a 'preestablished harmony'.

Later in life, his views matured to those that would have made him feel quite at home in early Greece:

I am convinced that we can discover by means of purely mathematical constructions the concepts and the laws connecting them with each other, which furnish the key to the understanding of natural phenomena. Experience may suggest the appropriate mathematical concepts, but they most certainly cannot be deduced from it. Experience remains, of course, the sole criterion of the physical utility of a mathematical construction. But the creative principle resides in mathematics. In a certain sense, therefore, I hold it true that pure thought can grasp reality, as the ancients dreamed.

Behind this change in attitude can be found an interesting change in Einstein's attitude to mathematics. In his early work on the special theory of relativity, Brownian motion, and also the photoelectric effect, for which his Nobel prize was awarded, we find that his style is to avoid complicated mathematics and stress simple physical arguments that get to the heart of the phenomenon in question. But his creation of the general theory of relativity introduced him to powerful mathematical formalism and the way in which the creations of the pure mathematicians can do more than merely describe the world. They can embody the very physical notions that one might otherwise struggle to impose upon a theory of Nature in a universal fashion. Impressed by the success of high-level mathematics in the formulation of the general theory of relativity in 1915, we find that Einstein's life-long quest for a unified field theory was dominated by the search for more general mathematical formalisms that could bring together the existing descriptions of gravity and electromagnetism. We find none of Einstein's compelling thought experiments and beautifully simple physical reasoning that lay at the heart of his early success. As the last quotation tells, he had become convinced that by pursuing mathematical formalisms alone, the compelling simplicity of a unified description of the world would become inescapable.

WHAT IS MATHEMATICS?

An unlucky accident has happened to the French mathematicians at Peru. It seems that they were shewing some French gallantry to the natives' wives, who have murdered their servants, destroyed their Instruments and burn't their papers, the Gentlemen escaping narrowly themselves. What an ugly article this will make in a journal.
— [letter to James Stirling (1740)]   COLIN MACLAURIN

At the end of the last century, a number of positions were laid down in response to this question of the nature and identity of mathematics. They were motivated by a number of contemporary problems concerning the scope of mathematics and the significance of logical paradoxes. They solidify into four simple alternative philosophies of mathematics.

The first, *formalism*, avoids any discussion of the meaning of mathematics by defining mathematics to be nothing more nor less than the set of all possible deductions from all possible sets of consistent axioms using all possible rules of inference. The resulting web of logical connections is what the early formalists took to encompass all mathematical truth. Any statement in the language of mathematics could be examined to discover whether or not it was a correct deduction from self-consistent axioms. No paradoxes could conceivably be deduced if the rules of inference were correctly implemented. Clearly, this rather claustrophobic picture of mathematics can offer us no help with our problem of why mathematics 'works'. It is just a logical game, like chess or 'go'. It does not *mean* anything. However, as is now fairly well known, this grandiose attempt to tie things up failed. It was shown first by Kurt Gödel that there must exist statements whose truth or falsity can never be demonstrated from the rules of deduction if they and the initial axioms are rich enough to include our familiar arithmetic of whole numbers. This we discussed from another point of view in Chapter 3. Hence, one cannot define mathematics in this straightforward formalistic manner, as we could, for example, define all possible games of noughts and crosses.

The second option that we have available is a philosophy of mathematics which I term *inventionism*. This regards mathematics as a purely human invention. Like music or literature, it is a product of the human mind. Mathematics is nothing more and nothing less than what mathematicians do. We invent it, we use it, but we do not discover it. No 'other world' of mathematical truths sits waiting to be uncovered. We have found mathematics to be the most useful mental scaffolding to erect in order to find our way around the fabric of the physical world. Reality is not intrinsically mathematical. Rather, it is only those aspects of reality amenable to mathematical description that we are any good at elucidating. Thus, it is argued that its effectiveness in describing the world is nothing more than a description, one that is effective because we have invented or selected mathematical tools as the ones that best do the job in each individual case. This view is most prevalent amongst economists, social scientists, and other consumers of mathematics who have to deal with very complex systems where symmetry plays no role or where events are the messy, haphazard outworkings of the process of natural selection. In many cases, these subjects focus attention upon the results of the organizing processes (or lack of them) that we discussed in an earlier chapter. They are far from the most pristine laws of Nature. This view looks at the mathematical dexterity of the human mind as an evolutionary effect, which goes some way towards explaining why there is a good match between our mental representation of the world and reality itself. Our brains are the outcome of some evolutionary history that has no preordained goal. But the most probable outcome of this history will be a mental apparatus for gathering, representing, and using information about the world in order to predict its future course, a representation that becomes more and more accurate in its reflection of the true underlying reality. A poor mental categorization of the physical world would have a low survival value in comparison with one that was accurate. Any creature that thought that here was there or before was after, who failed to recognize the process of cause and effect, would be less likely to survive and reproduce and so would become an increasingly minor contributor to the gene pool. This gives some historical credence to a realist picture of the world—up to a point. For there are parts of reality, like the world of elementary particles or cosmology, high-temperature superconductivity or quantum mechanics, of which we neither knew nor needed to know during the crucial evolutionary history that led to our mental facility. Perhaps these esoteric areas merely make complicated use of basic concepts whose faithful representation was honed by natural selection during our primitive past. The alternative that was proposed by Niels Bohr as an explanation for our struggle to come to terms with the interpretation of quantum complementarity is that there are concepts and areas of physical reality about which we have poor conceptual understanding precisely because the necessary ideas can have played no role in our evolutionary history. According to Bohr, our 'categories of thought', those mental filters of the sense data that we gather from the world,

> developed for orientation in our surroundings and for the organisation of human communities.

When we encounter events removed from everyday experience, he anticipated that there should arise difficulties of orienting ourselves in a domain of experience far from that to the description of which our means of expression are adapted.

Of course, one can extrapolate this approach even further to 'explain' all manner of mysteries, but one must take care to identify a precise neurophysiological adaptation in order that a degeneration into mere 'just-so' stories does not result. In many cases, a form of game-theoretic analysis is persuasive. For example, we might try to understand on evolutionary grounds why individuals tell the truth most of the time but not always. If everyone told the truth all the time, then the potential advantage to be gained by any deviant liar would be enormous. On the other hand, if everyone lies all the time, then society breaks down. In between, there would appear to exist a natural stable state in which most people tell the truth most of the time but lies are just common enough to prevent us from becoming the gullible victims of an inveterate liar.

For the inventionist, mathematics is one of our categories of thought and fundamental limitations upon its scope, like that discovered by Gödel, are associated with our categories of thought rather than with reality itself.

The next option as to the nature of mathematics is the realist or *Platonic* interpretation. Superficially the simplest, it maintains that mathematics really exists—'pi' really is in the sky—and mathematicians simply discover it. Mathematical truth exists independently of the existence of mathematicians. It is a form of objective universal truth. Thus the reason why mathematics is so successful in describing the way the world works is because the world *is* at root mathematical. Any limitations of mathematical reasoning, like those uncovered by Gödel, are thus not merely limitations on our mental categories but intrinsic properties of reality and hence limitations upon any attempt to understand the ultimate nature of the Universe.

On this interpretation, the Theory of Everything must be a mathematical theory. Of course, supporters of this view cite the successful mathematical description of the world as evidence in its support. But, even so, one might wonder why such elementary mathematical concepts are able to describe so much of the world. Perhaps we have not looked far enough into its structure to uncover very much of the truly deep and difficult structure. Certainly, the problems raised by the string picture of matter at its most elementary level have provoked its creators to claim that it has been discovered rather prematurely, long before our mathematical knowledge has matured sufficiently to cope with the questions that it raises. Certainly, it is a striking example of a physical theory that has found off-the-shelf mathematics insufficient for its purposes and has actually directed mathematicians into new and fruitful areas of pure mathematical enquiry.

Besides the traditional questions of where or what the Platonic world of perfect mathematical blueprints actually *is*, this view moves us towards a number of deep and fascinating questions. It elevates mathematics close to the status of God in traditional theology. Just take any work of medieval theology and alter the word 'God' to 'mathematics' wherever it appears and it makes pretty good sense. Mathematics is part of the world, and yet transcends it. It must exist before and after the Universe. In this respect, it is reminiscent of our discussion of the nature of time in earlier chapters. In the Newtonian world-view, both space and time were absolute and independent of the events played out upon them. Then the Einsteinian transformation of our concepts of space and time (whose radicalness is obscured by the fact that the concepts retained the same names) linked space and time to events going on within the Universe. Maybe a similar evolution of this interpretation of mathematics will emerge? Although at present mathematics seems to transcend the Universe because cosmologists think they can describe the Universe as a whole in terms of mathematics and use mathematics to study the process of creation and annihilation of universes, perhaps the nature of mathematics will become more closely associated with physically realizable processes like counting or computing?

Most scientists and mathematicians operate as if Platonism is true, regardless of whether they believe that it is. That is, they work as though there were an unknown realm of truth to be discovered.*

* An interesting, and somewhat subtle alternative perspective is offered by the so-called 'deflationist' philosophy of mathematics. This is a non-realist philosophy of mathematics that maintains that we cannot know, or even have grounds to believe in, the reality of the mathematical entities that the Platonist takes for granted. But, unlike other non-realist philosophies, it takes the successful application of mathematics to the real world seriously and attempts to explain it without assuming that mathematical claims are true (rather than merely being able to produce other mathematical claims) or that mathematical entities exist. It claims that the successful application of mathematics to the world requires only that it possess the property of *strong consistency*. A mathematical theory M is said to be strongly consistent if, when we add it to some consistent theory T that says nothing about mathematical entities, the resulting theory $T + M$ is also consistent. It is interesting that, whilst truth does not imply the property of strong consistency, the property of necessary truth does. In general, there is a circle of logical implications which connect theories which are true, necessarily true, consistent, and strongly consistent, which can be pictured as follows (with → meaning 'implies'):

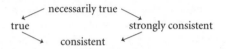

This approach tries to isolate the 'nice' feature that characterizes mathematical theories which make them useful. It stresses that truth is not sufficient to guarantee that a mathematical theory be 'nice' in this way; nor are all 'nice' theories necessarily true. When we apply mathematics to the physical world, all we require is the property of strong consistency, not the stronger property of truth.

Particle physicists are the most deeply Platonic because their entire sub-ject is built upon a belief that the deepest workings of the world are based upon symmetries. They examine symmetry after symmetry, confident in the expectation that the biggest and the best will have been employed in the grand scheme of things. But the Platonic philosophy is not so common as it was a hundred and fifty years ago, when the Victorians confidently filled their libraries with books bearing titles like *The Theory of Sound* or *Hydrodynamics*. Today the inventionists' nervousness about the existence of any such unique representation of things mathematical, independent of the mind of the mathe-matician, is reflected by more fashionable titles like *Mathematical Modelling of Sonic Phenomena* or *Concepts of Fluid Flow* which stress the subjective picture and non-uniqueness of the story within.

One dramatic consequence of the Platonic view is that life must exist in every sense because there exists a mathematical model of it. If we were to build a computer simulation of the evolution of a small part of the Universe, including, say, a planet like the Earth, then this model could in principle be refined to such an extent that it would include the evolution of sentient beings who would be self-aware. They would know of and communicate with other similar beings which arise within the simulation and could deduce the programming rules which they would designate as 'laws of Nature'. We know that such a program does exist simply because we have evolved and the sequence of events that led to us could in principle be simulated. But, because the program exists in principle, it can be argued that the sentient beings ne-cessarily exist in the only sense that is meaningful for them. Indeed, we could be part of such a simulation in the mind of God. Frank Tipler sums up the most radical interpretation of this type of computer ontological argument as follows:

A program sufficiently complex to contain observers *necessarily* exists. The idea is that all physical processes can be represented by a computer program. Thus, a sufficiently complex program can simulate the entire universe. In fact, if the simulation is perfect, then it would be by definition, completely indistinguish-able from the actual universe. Every person, and every person's every action in the actual universe, would have an exact analogue in the simulation. According to the simulated observations of the simulated people in the simulated uni-verse, they are real, they exist. We ourselves could be such a computer simu-lation. There would be no way to tell from inside the simulation that we were inside; the software cannot tell on which hardware it is being run. In fact, there is no reason for there to be any hardware; as Minsky puts it 'the uni-verse simply doesn't exist'. Thus, if a program—or more generally, a physical theory—which contains observers, exist mathematically, it necessarily exists physically, in the only reasonable sense of physical existence: observers observe themselves to exist.

The reader will have detected one dangerous bend in this argument—that every physical process can be simulated by a computer program. We do not know whether this is true; certainly, not every mathematical operation can be simulated in this way. There exist 'true' theorems that cannot be deduced step by step after the fashion of a computer program. We shall have a lot more to say about this question in a little while. We have raised the question here because it leads us naturally into the third of our possible interpretations of mathematics.

Constructivism was conceived in the climate of uncertainty at the end of the nineteenth century created by the logical paradoxes of set theory and the strange Cantorian properties of infinite sets we introduced in Chapter 3. Sensing that mathematics might be led into serious error and contradiction by the manipulation of concepts, like infinity, about which we have no concrete experience, it was proposed by some that we should adopt a conservative stance which defined mathematics to include only those statements that could be deduced in a finite sequence of step-by-step constructions starting from the natural numbers, which were assumed as God-given and fundamental. At first, this sounds little more than a time-consuming piece of bureaucracy, but it turns out to have the most dramatic consequences for the whole scope and meaning of mathematics.

The confinement of logical argument to the constructivists' dicta removes such familiar devices as the argument from contradiction (the so-called *reductio ad absurdum*), wherein one assumes some statement to be true and from that assumption proceeds to deduce a logical contradiction and hence to a conclusion that the original assumption must have been false. If the constructivist philosophy is adopted, then the content of mathematics is considerably reduced. The results of such a descoping are significant for the scientist also. Indeed, we would have to relinquish such famous deductions as the 'singularity theorems' of general relativity which specify the conditions which, when satisfied by the structure of a Universe and its material content, suffice to indicate the existence of a past moment when the laws of physics must have broken down—a singularity which we have come to call the 'Big Bang'. For these theorems do not construct this past moment explicitly, rather they use the device of *reductio ad absurdum* to show that its non-existence would result in a logical contradiction. The important lesson that we learn here is that the notion of what is 'true' about the Universe appears to depend upon our philosophy of mathematics. This is the down-side of living in a world that is so evidently mathematical.

Clearly, the constructivist is a close cousin of the operationalist philosopher who defines things by the process by which they can be implemented or constructed. The most interesting physical quantity in this respect is 'time', which, if defined in terms of the process by which it is recorded, leads to the possibility that we might judge the Universe to have had an infinite past age if there are fundamental measures of things happening in the Universe which slow down as one goes further and further back into the past.

The constructivist philosophy leads naturally to the concept of computers, for the step-by-step construction of mathematical statements is what computers do. The essence of all practical computers amounts merely to an ability to read a string of integers and transform them into another string of integers. This ability, despite its superficial triviality, is all that is required for the activities of the world's most powerful computers. Their impressiveness as calculating engines resides in the speed with which they can carry out such operations, together with their ability sometimes to carry out several such operations simultaneously. This capability at the heart of all calculating devices is called that of a *Turing machine*, named after the English mathematician Alan Turing. The term 'Turing machine' is used to characterize the capability of any step-by-step logical device.

Originally it had been hoped that such a hypothetical device might be able to carry out any mathematical operation and thus enable all the decidable truths of mathematics to be catalogued mechanically. Alan Turing, Alonzo Church, and Emil Post were the first to show that this cannot be so. There exist mathematical operations—so-called *non-computable functions*—which cannot be conducted by any Turing machine. Of course, there also exist mathematical operations which the machine can carry out in a step-by-step manner, but which take millions of years to complete using the fastest available machines. Such operations are for all practical purposes non-computable and in fact form the basis of many modern forms of encryption. However, they are qualitatively different from non-computable functions in that the latter would require an infinite period of time for any Turing machine to carry it out: the Turing machine would never reach the stage of printing out the final result.

If true, the constructivist picture of mathematics has a number of illuminating things to reveal about the mathematical Universe. It sets in a new light the question of why mathematics is so unreasonably effective in describing the real world. This effectiveness is equivalent to the fact that so many simple mathematical operations are computable operations in the Turing sense.

Computable functions are mathematical operations that can be simulated by a real device, an artefact of the physical world made out of elementary particles constrained by the laws of Nature. Conversely, the fact that real devices or 'natural phenomena' are well described by simple mathematical functions is equivalent to the fact that so many of those functions are computable. If all the simple functions of mathematics were non-computable, then we would not find mathematics a useful vehicle for describing the world. We could have non-constructive theorems or truths about the world, but little of practical applicability.

Structuralism is the last of our possible interpretations of mathematics. It is the weakest of all the definitions we have seen, but it the most straightforward for the outsider to appreciate and it avoids extreme metaphysical steps. It asks us to take mathematics to be the catalogue of all possible patterns. Some of those patterns can be found on our walls or in the flowers that grow in the garden; others are in sequences of movements of the heavenly bodies; while others are to be found in computational steps or inside our heads. When mathematics is viewed in this light its utility is no longer a mystery. The Universe must contain some order—some patterns—if life is to exist and evolve within it. Mathematics is just the description of these inevitable patterns. However, the central mystery is just shifted sideways a little. Why are such *simple* patterns so far-reaching? It could have been that the Universe was structured by patterns that were uncomputable in Turing's sense and so mathematics would describe the Universe but be of no practical use in predicting the future or understanding the present. Our Universe is extraordinarily compressible in the sense that its diversity and complexity can be reduced to the outworkings of a very small number of fairly simple patterns. It is the goal of the physicists' search for a Theory of Everything to reduce that number to one in a well defined sense.

MATHEMATICS AND PHYSICS: AN ETERNAL GOLDEN BRAID

Do not infest your mind with beating on
The strangeness of this business.
— WILLIAM SHAKESPEARE

The striking symbiosis of mathematics and physics admits of examples that span many centuries. This relationship also possesses a surprising symmetry: there are examples where old mathematics finds itself tailor-made to further the description of the physical and there are examples where the desire to advance our understanding of the physical world has led to the creation of new mathematics which mathematicians subsequently pursue for its own sake. Let us consider a few notable examples in each category. First, some examples where mathematical ideas had been developed for their own sake by mathematicians impressed by the symmetry, intrinsic logic and generality of the concepts involved but which were then found to provide the perfect vehicle for the description and elucidation of new aspects of Nature.

Conics

Apollonius of Perga lived from about 262 to 200 BC, and was a contemporary of Archimedes. He learnt his mathematics in the school founded by Euclid's successors and, despite the fact that most of his original works have been lost, he is established as one of the greatest mathematicians of antiquity by virtue of his work on conics. This lays down all the geometrical and algebraic properties of ellipses, parabolae, and hyperbolae. Without these investigations, Kepler would have lacked the mathematical descriptions required for the theory of planetary motions which he formulated in 1609. Later, Newton's derivation of Kepler's laws of planetary motion from his own inverse-square law of gravitational force displayed the full physical significance of parabolic, hyperbolic, and elliptic curves in the description of the orbits of bodies moving under attractive force-fields like gravitation.

Riemannian geometry and tensors

The development of non-Euclidean geometry as a branch of pure mathematics by Riemann in the nineteenth century, and the study of mathematical objects called tensors was a godsend to the development of twentieth-century physics. Tensors are defined by the fact that their constituent pieces change in a very particular fashion when their coordinate labels are altered in completely arbitrary ways. This esoteric mathematical machinery proved to be precisely what was required by Einstein in his formulation of the general theory of relativity. Non-Euclidean geometry described the distortion of space and time in the presence of mass-energy, while the behaviour of tensors ensured that any law of Nature written in tensor language would automatically retain the same form no matter what the state of motion of the observer. Indeed, Einstein was rather fortunate in that his long-time friend, the pure mathematician Marcel Grossmann, was able to introduce him to these mathematical tools. Had they not already existed Einstein could not have formulated the general theory of relativity.

Groups

We have repeatedly stressed the over-arching role of symmetry in modern physics. The systematic study of symmetry falls under the heading of 'group theory' for the mathematician. This subject largely was created during the mid-nineteenth century, again without physical motivation; it split into the study of finite groups, which describe particular discrete changes, like rotations, and those which describe continuous transformations. The latter were studied in overwhelming detail by the Norwegian Sophus Lie. Indeed, the power and profundity of these nineteenth-century developments, and the manner in which they shed light upon what appeared totally different areas of mathematics, led Henri Poincaré to claim that groups are 'all of mathematics'. Yet, at that time, there was no evident connection with any problems of physics and in 1900 Sir James Jeans, when commenting to a colleague upon the areas of mathematics that were most fruitful for the physicist to know, asserted that

we may as well cut out group theory, that is a subject which will never be of any use in physics.

On the contrary, it is the systematic classification of symmetry and its canonization into the subject of group theory which forms the basis of so much of modern fundamental physics. Nature likes symmetry and so groups form a fundamental part of its description.

Hilbert spaces

There are two great physical theories of twentieth-century physics. The first, general relativity, as we have just seen, owes its creation to the availability of a large body of non-Euclidean geometry and tensor calculus. The second, quantum mechanics, is no less indebted to the mathematicians. In this case, it was David Hilbert who was the unwitting midwife. Hilbert created the idea of constructing infinite-dimensional versions of Euclidean space. These are now called Hilbert spaces. A Hilbert space is a space whose points are in one-to-one correspondence with a collection of mathematical operations of a certain type. These spaces form the basis for the mathematical formulation of the theory of quantum mechanics and most of the modern theories of elementary-particle interactions. Much of this abstract mathematics was invented in the early years of the twentieth century and was available for the physicists to exploit twenty years later when the quantum revolution led by Bohr, Heisenberg, and Dirac was fully formalized.

Complex manifolds

The recent interest in superstring theories as candidates for a Theory of Everything has sent physicists rushing to their mathematics books once again. On
this occasion, they required ideas about the structure of complex manifolds
and other equally esoteric pure mathematical generalizations of more familiar
concepts. However, on this occasion, the books were often empty. For the
first time in the recent history of the subject, physicists have found that off-
the-shelf mathematics is insufficient for their purposes and mathematicians
have set to work to extend their subject in the directions that physicists
require.

The parts of mathematics required to pursue the concept of strings as the
most fundamental entities of the physical world are at the frontier of mathematical knowledge. Few physicists are equipped to understand them fully and
the style of investigation presented by the mathematician is often exasperating
to the physicist. The physicist wants to understand things in a fashion that
will enable him to use them in his work. This requires the development of
a good intuitive feel for the ideas in question and is often most effectively
arrived at by devising simple examples of the abstract concepts involved. That
is, the physicist likes to learn from particular illustrations of a general abstract
concept. The mathematician, on the other hand, often eschews the particular
in pursuit of the most abstract and general formulation possible. Although the
mathematician may think from, or through, particular concrete examples in
coming to appreciate the likely truth of very general statements, he will hide all
those intuitive steps when he comes to present the conclusions of his thinking.
As a result, the pure mathematical research literature is virtually impenetrable
to outsiders. It presents the results of research as a hierarchy of definitions,
theorems and proofs after the manner of Euclid; this minimizes unnecessary
words but very effectively disguises the natural train of thought that led to
the original results. To a great extent, this unfortunate trait was encouraged
by the Bourbaki project. Bourbaki is a pseudonym for an evolving group of
French mathematicians, who in the last fifty years have co-authored a series
of monographs about the fundamental 'structures' of mathematics. They
represent the last hopes of the formalists: axiomatics, rigour, and elegance
prevail; diagrams, examples, and the particular are excluded. Although the
score or more Bourbaki volumes have not presented new mathematical results
they have presented known areas of the subject in new and more abstract
ways. They are the ultimate textbooks for the cognoscenti. Even within the
mathematical ranks, Bourbaki has vociferous critics of its 'scholasticism' and
'hyperaxiomatics', but one of its supporters, Laurent Schwarz, tries to justify

its approach and the way in which it contrasts with that of the inventors of new ideas in the following way:

> Scientific minds are essentially of two types, neither of which is to be considered superior to the other. There are those who like fine detail, and those who are only interested in grand generalities...In the development of a mathematical theory, the ground is generally broken by scientists of the 'detailed' school, who treat problems by new methods, formulate the important issues that must be settled, and tenaciously seek solutions, however great the difficulty. Once their task is accomplished, the ideas of the scientists with a penchant for generality come into play. They sort and sift, retaining only material vital for the future of mathematics. Their work is pedagogic rather than creative but nevertheless as essential and difficult as that of thinkers in the alternative category...Bourbaki belongs to the 'general' school of thought.

However, the mainstream of mathematics has begun to move away from the high-ground of extreme formalism back to the study of particular problems, notably those involving chaotic non-linear phenomena, and to seek motivation from the natural world. This is a return to a distinguished tradition for, just as there are examples of old mathematics proving appropriate to induct us into new physics, so there are complementary examples where our study of the physical world has motivated the invention of new mathematics. The contemplation of continuous motion by Newton and Leibniz and their desire to give meaning to the notion of an instantaneous rate of change of a quantity led to the creation of the calculus. The work of Jean-Baptiste Fourier on the series of trigonometric curves now known as 'Fourier series' arose from the study of heat flow and optics. In the twentieth century, the consideration of impulsive forces led to the invention of new types of mathematical entity called 'generalized functions'. They were used most powerfully by Paul Dirac in his formulation of quantum mechanics and then axiomatized and generalized into a subset of pure mathematics. This evolution was memorably summarized by James Lighthill, who wrote the first expository textbook on the subject, when giving credit to Dirac, Laurent Schwarz (who provided the rigorous pure mathematical justification of the notions used intuitively by Dirac), and George Temple (who showed how Schwarz's logical edifice could be simply explained to those who wanted to use it). Lighthill dedicates his work thus:

> To Paul Dirac, who saw that it must be true. To Laurent Schwarz, who proved it. And to George Temple, who showed how simple it all could be.

In recent years, this trend towards specific applications has been perpetuated by the creation of a large body of dynamical systems theory, and most notably the concept of a strange attractor, as a result of the quest to describe turbulent fluid motions. The growing interest in the description of chaotic change which is characterized by the very rapid escalation of any error in its exact description as time passes has led to a completely new philosophy with regard to the mathematical description of phenomena. Instead of seeking more and more accurate mathematical equations to describe a given phenomenon, one searches for those properties which are possessed by almost every possible equation governing change. Such 'generic' properties, as they are called, can therefore be relied upon to manifest themselves in phenomena that do not possess very special properties. It is this class of probable phenomena that are most likely to be found in practice.

Finally, we could return to the situation of strings and complex manifolds. This area of physics, although as yet lacking any experimentally testable outcome, has pointed the way towards new types of mathematical structure and its foremost proponents hover on the brink of becoming pure mathematicians in all but name. If superstring theory does manage to produce some observable predictions in the not too distant future, then we may witness the interesting spectacle of pure mathematics receiving marching orders from experimental physics once again. Then again, if strings were definitively excluded as a description of the physical world of elementary particles, the pure mathematicians would remain interested in their mathematical structure alone. Like a genie, once released from its pot, it is not so easy to entice it back.

THE INTELLIGIBILITY OF THE WORLD

They say it takes three generations to learn how to cut a diamond, a lifetime to learn how to make a watch and that only three people in the entire world ever fully comprehended Einstein's Theory of Relativity. But football coaches to a man are convinced that none of the above is comparable in complexity to playing quarterback in the NFL. I mean, watches don't mix up defenses on you, diamonds don't blitz and Einstein had all day to throw. $E = mc^2$ doesn't rotate coverages.
— LOS ANGELES TIMES SPORTS REPORT

One of the most striking features of the world is that its laws seem simple whilst the plethora of states and situations it manifests are extraordinarily complicated. To reconcile this disparity, we must redraw a sharp distinction that we made in Chapter 6 between the laws of Nature and the outcomes of those laws, between the equations of physics and their solutions. The fact that the outcomes of the laws of Nature need not possess the symmetries of the underlying laws makes science difficult and teaches us why the complicated collective structures we find in Nature can be the outcomes of very simple laws of change and invariance. But, however necessary an appreciation of this point might be, it is far from sufficient to make sense of the physical world.

On the face of it, we might imagine that it would be far easier for the world to be an unintelligible chaos than the relatively coherent cosmos that the scientist delights in unravelling. What are the features of the world which play important roles in rendering it intelligible to us? Here is a catalogue of those aspects which appear to play a subtle but vital role in the intelligibility of Nature.

Linearity

Linear problems are easy problems. They are those problems where the sum or difference of any two particular solutions is also a solution. If L is a linear operation and its action upon a quantity A produces the result a, whilst its operation upon B produces the result b, then the result of the operation of L upon A plus B will be a plus b. Thus, if a situation is linear or dominated by influences that are linear, it will be possible to piece together a picture of its whole behaviour by examining it in small pieces. The whole will be composed of the sum of its parts. Fortunately for the physicist, a large part of the world is linear in this sense. In this part of the world, one can make small errors in determining the behaviour of things at one time and those errors will only be amplified very slowly as the world changes in time. Linear phenomena are thus amenable to very accurate mathematical modelling. The output of a linear operation varies steadily and smoothly with any change in its input. Non-linear problems are none of these things. They amplify errors so rapidly that an infinitesimal uncertainty in the present state of the system can render any future prediction of its state worthless after a very short period of time. Their outputs respond in discontinuous and unpredictable ways to very small changes in their inputs. Particular local behaviours cannot be added together to build up a global one: a holistic approach is required in which the system is considered as a whole. We are familiar with many complicated problems of this sort: the surge of water from a tap, the development of a complex economy, human societies, the behaviour of weather systems—their whole is more than the sum of their parts. Yet our education and intuition is dominated by the linear examples because they are simple. Educators display the solutions of linear equations and textbook writers present the study of linear phenomena because they are the only examples that can be solved easily: the only phenomena that admit of a ready and complete understanding. Many social scientists who seek mathematical models of social behaviour invariably look to linear models because they are the simplest and the only sort they have been taught about. Yet the simplest imaginable non-linear equations exhibit behaviour of unsuspected depth and subtlety which is, for all practical purposes, completely unpredictable.

Despite the ubiquity of non-linearity and complexity, the fundamental laws of Nature often give rise to phenomena that are linear. Thus, if we have a physical phenomenon that can be described by the action of a mathematical operation f upon an input x, which we denote by $f(x)$, then in general we can express this as a series of the form

$$f(x) = f_0 + x f_1 + x^2 f_2 + \cdots,$$

Where the series could go on forever. If $f(x)$ is a linear phenomenon, then it can be very accurately approximated by the first two terms of the series on the right-hand side of the equation; the remaining terms are either all zero or else they diminish in size so rapidly as one goes from one term to the next that their contribution is negligible. Fortunately, most physical phenomena possess this property. It is crucial in rendering the world intelligible to us and it is closely associated with other aspects of reality, the most notable of which we shall highlight next.

Locality

The hallmark of the entire non-quantum world is that things occurring here and now are directly caused by events that occurred immediately nearby in space and time: this property we term 'locality' to reflect the fact that it is the most local events that exert the predominant influence upon us. Usually, linearity is necessary for a law of Nature to possess this property, although linearity is not sufficient to guarantee it. The fundamental forces of Nature, like gravity, diminish in strength with increasing distance away from their source at a rate that ensures that the total effect at any point is dominated by the nearby sources rather than those on the other side of the Universe. Were the situation reversed, then the world would be erratically dominated by imperceptible influences at the farthest reaches of the Universe and our chances of beginning to understand it rendered pretty slender. Interestingly, the number of dimensions of space which we experience in the large plays an important role in ensuring this state of affairs. It also ensures that wave phenomena behave in a coherent fashion. Were there four dimensions of space, then simple waves would not travel at one speed in free space, and hence we would simultaneously receive waves that were emitted at different times. Moreover, in any world but one having three large dimensions of space, waves would become distorted as they travelled. Such reverberation and distortion would render any high-fidelity signalling impossible. Since so much of the physical universe, from brain waves to quantum waves, relies upon travelling waves we appreciate the key role played by the dimensionality of our space in rendering its contents intelligible to us.

Not every natural phenomenon possesses the property of locality. When we look at the quantum world of elementary particles, we discover that the world is non-local. This is the import of Bell's famous theorem. It reveals to us something of the ambiguity between observer and observed that arises when we enter the quantum world of the very small, where the influence of the act of observation upon what is being observed is invariably significant. In our everyday experience, this quantum ambiguity is never evident. We confidently hold such notions like position or speed to be well defined, unambiguous, and independent of who is using them. But the fact that our present-day Universe admits such definiteness is something of a mystery. As we look way back into the first instants of the Big Bang, we find the quantum world that we described in Chapter 3. From that state, where like effects do not follow from like causes, there must somehow emerge a world resembling our own, where the results of most observations are definite. This is by no means inevitable and may require the Universe to have emerged from a rather special primeval state.

The local–global connection

The helpful presence of linearity and locality in the world of everyday observation and experience was essential for the beginning of our understanding of the world. Such an understanding begins locally and finds local causes for local effects. But what must the world be like in order that we can piece together a global description from the local. In some sense, the global picture of the Universe must be built out of many copies of its local structure. Equivalently, there must exist some invariances of the world as we change the locations in space and in time of all its most elementary entities so that the most basic fabric of reality is universal rather than dependent upon parochial things. Particle physicists have discovered that the world is mysteriously structured in such a fashion and the local gauge theories which we introduced earlier in Chapter 4 bear witness to the power of this local–global connection. The requirement that there exist this natural correspondence between the local and the global structure is found to require the existence of the forces of Nature that we see. We do not mean this in a teleological sense. It is merely a reflection of the interwoven consistency and economy of the natural structure of things. The forces of Nature are not required as an additional *ad hoc* ingredient.

When we look in greater depth at the mathematical structures that are most effective in the description of the world, and indeed why there can be such effective structures, we find a situation of great subtlety. We find the presence of mathematical operations, like the expansion of a function in a series shown above or the 'implicit function theorem', which guarantees that, if some quantity is completely determined by the values of two variables x and y and is found to be a constant, then y can always be expressed as some function of the variable x alone. These two mathematical properties both define restrictions upon what local information about the world can be deduced from global (or large-scale) information. By applying these local restrictions to themselves over and over again in an iterative manner, we can build up increasingly global information about a mathematical world. By contrast, there exist examples of the converse. Stokes' famous integral theorem and the process of analytic continuation, familiar to undergraduates, are both examples wherein restrictions are defined for the transit from local to global information. They exemplify one of the goals of the human investigation of Nature: to extend our knowledge of the world from the local domain, to which we have direct access, to the wider scale, of which we are as yet ignorant. Stokes' theorem alone does not permit such an extension to be made unambiguously. It leaves an undetermined constant quantity undetermined at the end of the extension process. The power of gauge theories in physics derives from their ability to remove this arbitrariness in the extension process and determine the unknown constant uniquely through the imposition of symmetry and invariance upon the extension process.

All the best physical theories are associated with equations which allow the continuation of data defined at present into the future, and hence allow prediction. But this situation requires space and time to possess a rather particular type of mathematical property which we shall call 'natural structure'. Other theories, like those describing statistical or probabilistic outcomes, which attempt to use mathematics for prediction, often fail to possess a mathematical substratum with a 'natural structure' of this sort, and so there is no guarantee that its future states are smooth continuations of its present ones.

One feature of the elementary-particle world, which is totally unexpected when compared with our experience of everyday things, is the fact that elementary particles come in populations of universally identical particles. Every electron that we have encountered, whether it comes from outer space or a laboratory experiment, is found to be identical. All have the same electric charge, the same spin, the same mass, to the accuracy of measurement. They all behave in the same way in interaction with other particles. Nor is this fidelity confined to electrons: it extends to all the populations of elementary particles, from the quarks and leptons to the exchange particles that mediate the four fundamental forces of Nature. We do not know why particles are identical in this way. We could imagine a world in which electrons were like footballs—everyone slightly different to all the others. The result would be an unintelligible world.

In fact, even in a world populated by collections of identical elementary particles, there would not exist populations of identical larger systems, composed of systems of those particles unless energy is quantized in some way. Although the uncertainties introduced by the quantum picture of reality are often stressed, this same quantum structure is absolutely vital for the stability, consistency, and intelligibility of the physical world. In a Newtonian world, all physical quantities, like energy and spin, can take on *any* values whatsoever. They range over the entire continuum of numbers. Hence, if one were to form a 'Newtonian hydrogen atom' by setting an electron in circular orbit around a single proton then the electron could move in a closed orbit of any radius because it could possess any orbital speed. As a result, every pair of electrons and protons that came together would be different. The electrons would find themselves in some randomly different orbit. The chemical properties of each of the atoms would be different and their sizes would be different. Even if one were to create an initial population in which the electrons' speeds were the same and the radii of their orbits identical, they would each drift away from their starting state in differing ways as they suffered the buffetings of radiation and other particles. There could not exist a well-defined element called hydrogen with universal properties, even if there existed universal populations of identical electrons and protons. Quantum mechanics shows us why there are identical collective structures. The quantization of energy allows it to come only in discrete packets, and so when an electron and a proton come together there is a single state for them to reside in. The same configuration arises for every pair of electrons and protons that you care to choose. This universal state is what we call the hydrogen atom. Moreover, once it exists, its properties do not drift because of the plethora of tiny perturbations from other particles. In order to change the orbit of the electron around the proton, it has to be hit by a sizeable perturbation that is sufficient to change its energy by a whole quantum packet. Thus the quantization of energy lies at the root of the repeatability of structure in the physical world and the high fidelity of all identical phenomena in the atomic world. Without the quantum ambiguity of the microscopic world the macroscopic world would not be intelligible, nor indeed would there be intelligences to take cognisance of any such a totally heterodox non-quantum reality.

Symmetries are small

The possibilities open to an elementary particle of Nature amount to every-thing compatible with the maintenance of some symmetry. The preservation of some global pattern in the face of all the local freedoms to change is equiva-lent to a conservation law of Nature and all laws of change can be re-expressed in terms of the invariance of some quantity. The particular patterns that are generated arise from the concatenation of a finite number of ingredients. For example, a collection of patterns might be created from a combination of rotations and straight-line movements in space. The greater the number of distinct operations, or generators, that comprise the overall collection of patterns that can be generated, so the greater will be the number of patterns. If this number is very large, then for all practical purposes there will be no discernible symmetry at all. The generators of the symmetries that dictate the interactions that can occur between elementary particles are equivalent to the particles that mediate the force of Nature in question. Hence, the intelligibility of the world relies upon the fact that there are relatively few types of elementary particle. They are numbered in tens rather than in thousands or millions.

There exists one further connection between the population of the elementary-particle world and the overall simplicity of Nature. The unifi-cation of the forces of Nature that we have discussed in earlier chapters relies upon the property of 'asymptotic freedom' that is manifested by the strong force between particles like quarks and gluons which carry the colour charge. This means that, as the energy of the interaction between the particles increases, the strength of their interaction decreases, so that 'asymptotically' there would remain no interaction at all and the particles would be free. It is this property which allows the disparate forces of Nature that we witness at low energies to become unified at high energies. However, this feature would not arise if there existed too many elementary particles. For example, if there were eight types of neutrino rather than the three that experiments tell us there are, then interactions would get stronger rather than weaker as we go to higher energies, and the world would become intractably complicated as we scrutinize over finer and finer dimensions of the microscopic world.

This list of properties which may be necessary for the intelligibility of the world is not intended to be exhaustive, merely illustrative. It will not have escaped the reader's attention that many of the properties we have unveiled are also likely to be necessary properties for the existence of complex stable systems in the Universe, a subset of which we would call 'living'. We can conceive of universes where living observers (not necessarily resembling ourselves) could not exist and, perhaps unexpectedly, we find that there is an intimate connection between the most basic elements of the Universe's fabric and the conditions required for the evolution of life to have a probability that is distinguishable from zero.

ALGORITHMIC COMPRESSIBILITY RIDES AGAIN

The brain is a wonderful organ; it starts working the moment you get up in the morning, and does not stop until you get to the office.
— ROBERT FROST

At root, all the necessary conditions for the intelligibility of the world that we have been discussing amount to conditions which enable us to make sense out of what would otherwise be an intractable chaos. 'Making sense' of things amounts to cutting them down to size, ordering them, finding regularities, common factors, and simple recurrences which tell us why things are as they are and how they are going to be in the future. This we should now recognize as the quest for algorithmic compressibility, which we introduced in our opening chapter.

In practice, the intelligibility of the world amounts to the fact that we find it to be algorithmically compressible. We can replace sequences of facts and observational data by abbreviated statements which contain the same information content. These abbreviations we often call 'laws of Nature'. If the world were not algorithmically compressible, then there would exist no simple laws of Nature. Instead of using the law of gravitation to compute the orbits of the planets at whatever time in history we want to know them, we would have to keep precise records of the positions of the planets at all past times; yet this would still not help us one iota in predicting where they would be at any time in the future. The world is potentially and actually intelligible because at some level it is extensively algorithmically compressible. At root, this is why mathematics can work as a description of the physical world. It is the most expedient language that we have found in which to express those algorithmic compressions.

We know that the world is not totally algorithmically compressible. There exist particular chaotic processes that are not algorithmically compressible, just as there exist mathematical operations that are non-computable. And it is this glimpse of randomness that gives us some inkling of what a totally incompressible world would look like. Its scientists would be librarians rather than mathematicians, cataloguing fact after unrelated fact.

We see science as the search for algorithmic compressions of the world of experience and the search for a single all-encompassing Theory of Everything as the ultimate expression of some scientists' deeply held faith that the essential structure of the Universe as a whole can be algorithmically compressed. But we recognize that the human mind plays a non-trivial role in this evaluation. Inextricably linked to the apparent algorithmic compressibility of the world is the ability of the human mind to carry out compressions. Our minds have evolved out of the elements of the physical world and have been honed, at least partially, towards their present state by the perpetual process of natural selection. Their effectiveness as sensors of the environment, and their survival value, are obviously related to their abilities as algorithmic compressors. The more efficiently they can store and codify an organism's experience of the natural world, so the more effectively can that organism counter the dangers that an otherwise unpredictable environment presents. In our most recent phase of history as *Homo sapiens*, this ability has attained new levels of sophistication. We are able to think about thinking itself. Instead of merely learning from experience as part of the evolutionary process, we have sufficient mental capacity to be able to simulate or imagine the likely results of our actions. In this mode, our minds are generating simulations of past experience embedded in new situations. But to do this effectively requires the brain to be rather finely balanced. It is obvious that mental capacity must be above some threshold in order to achieve effective algorithmic compression. Our senses have to be sensitive enough to gather a significant amount of information from the environment. But it is understandable why we have not become too good at this. If our senses were so heightened that we gathered every piece of information possible about the things that we saw or heard—all the minutae of atomic arrangements—then our minds would be overloaded with information. Processing would be slower, reaction times longer, and all sorts of additional circuitry would be required to sift information into pictures of different levels of intensity and depth.

The fact that our minds are not too ambitious in their information gathering and processing abilities means that the brain would effect an algorithmic compression upon the Universe whether or not it were intrinsically so compressible. In practice, the brain does this by truncation. Our unaided senses are only able to take in at most a certain quantity of information about the world down to some level of resolution and sensitivity. Even when we enlist the aid of artificial sensors like telescopes and microscopes to enlarge the range of our facilities, there are still fundamental limits to the extent of that extension. Often this truncation process becomes rather formalized as a branch of applied science in itself. A good example is statistics. When we study a large or very complicated phenomenon, we might try to algorithmically compress the information available by sampling it in some selective way. Thus, pollsters trying to predict public opinion before a general election should ask every individual in the country who they will vote for. In practice, they ask a representative subset of the population and invariably produce a startlingly good prediction of the results of the full election.

CONTINUITY—A BRIDGE TOO FAR?

Observers will tend to be computationally equivalent to the systems they observe—with the inevitable consequence that they will consider the behaviour of such systems complex.
—STEPHEN WOLFRAM

The development of modern physics and its mathematical description of the Universe has followed a path laid down first by Isaac Newton more than 300 years ago. It assumes that space and time are continuous qualities which are best described by mathematical operations controlled by differential equations. A continuous variation is one that is illustrated by a curve which can be drawn without taking your pencil off the page while you are drawing it. All the equations which describe the known laws of physics are differential equations and the Theory of Everything is expected to be described by differential equations as well. They are 'machines' for predicting the future given the present. Looked at more critically, they make a number of very powerful assumptions about the nature of the Universe. If we look at all the ways in which one collection of all possible numbers can be changed into another list of all possible numbers then the assumption that the transformations are continuous ones reduces the number of possibilities *infinitely*. A world that permits discontinuous changes as well as continuous ones is infinitely more diverse in the range of possibilities open to it in how it can change from place to place and time to time. In practice, we cannot tell whether space and time are really grainy and discontinuous because we can't examine them down to arbitrarily small distances and intervals. So, physicists have retained their liking for differential equations and assumed that the world is a continuum all the way down to the smallest imaginable dimensions of space and time.

In the last few years, Stephen Wolfram has argued for a thoroughgoing exploration of the alternative—that the world is ultimately discontinuous and we have just got into a habit of describing it continuously because mathematical physics was invented before computers. Wolfram used his *Mathematica* computer language to investigate all the possible rules governing discontinuous changes in which the change in the state of the system at one location is determined by the neighbouring states. *Mathematica* doesn't just crunch numbers like a big calculator: it manipulates symbols and does algebra. At first this seems a hopelessly complicated task. Although the number of possible transformation rules of this sort is finite, one might expect that the behaviours that result would be fantastically diverse and peculiar to the rule that is adopted. Wolfram used the considerable computing power at his disposal to investigate all these possible rules systematically.

The results of this ambitious search through the catalogue of discontinuous changes were a surprise to everyone. Despite the scope for almost anything to happen as the rules were changed, there were only ever four types of behaviour emerging. Either the pattern quickly dies out, oscillates back and forth between one or more simple states, expands to become unboundedly large, or it creates a random state of maximum complexity.

One might imagine that by introducing more complicated rules, even using rules that use averaged information from many neighbouring members of the present generation to evolve the next step for each location, more exotic possibilities might arise. Wolfram investigated many of these alternatives and many more general systems of evolution, and the conclusion he drew was striking: no significantly more complex behaviour emerged than was found in the first simple catalogue listed above. When he increased the number of dimensions of space so a location had more neighbours and still more possible rules of change, then exactly the same conclusion emerged. No essential increase in complexity occurs.

All this is rather unexpected. It is as if there is a maximum level of complexity that can exist in computational systems and this maximum is very rapidly and universally achieved, even with the simplest possible set of rules. This raises a number of new questions for the biological study of how complexity arises, and whether natural selection alone provides us with the full story about the origins of complexity in Nature. Wolfram's studies suggest that there is an easily attained maximum complexity in systems that follow simple rules of change and all the complex systems we see in Nature (including ourselves) are equivalent in some basic sense by virtue of having achieved this maximum available complexity.

This leads to the idea of a 'Principle of Computational Equivalence' that expects every computer—including natural ones like ourselves—to have the same maximum complexity level as the environment around them because it is so easily achieved. Hence, conscious thinking computers will tend to judge the systems they study as 'complex.' There is no vastly higher level of complexity for their minds to have achieved over and above that of the world around them. Nor does it matter whether the conscious computers are like us. They could be nanoscopic beings or huge extended minds, their intrinsic complexity will be qualitatively similar.

What is also surprising about this, if true, is that we might have expected the human mind to be vastly more complex computationally than rather simple sets of rules for changing one set of 0s and 1s into another set. But this need not be the case. The achievable level of complexity is already saturated by a game with simple rules, like John Conway's Game of Life. There are also possible implications for the explanation of complexity by means of evolution by natural selection. Generally, it is believed that the evolution of great complexity requires considerable 'effort' and time by the evolutionary process. But it could be that a significant component of any observed complexity arises as a mere by-product of simple rules without pressure from natural selection. Wolfram argues that with regard to the creation of biological complexity

> my strong suspicion is that the main effect of natural selection is almost exactly the opposite: it tends to make biological systems avoid complexity, and be more like systems in engineering... Whenever natural selection is an important determining factor, I suspect that one will inevitably see many of the same simplifying features as in systems created through engineering. And only when natural selection is not crucial, therefore, will biological systems be able to exhibit the same level of complexity that one observes for example in many systems in physics.

There is one awkward fact for this ambitious scenario. The aim was to show that it was possible to replace the descriptions of physical laws based on continuous differential equations by discontinuous models, using difference equations. The latter are much easier to deal with on a computer and, in reality, when you want to solve a differential equation approximately on a computer, the first step is to convert it into a difference equation. However, there is one essential part of physics that seems to resist replacement in this way. Quantum theory is intrinsically non-local; that is, it cannot be replaced by a model in which what is observed at one location is only determined by the states of immediately neighbouring locations: there are mysterious long-range entanglements in the quantum Universe.

THE SECRET OF THE UNIVERSE

This principle is so perfectly general that no particular application of it is possible.
— GEORGE POLYA

We have learnt that it is natural to describe a sequence as random if there exists no possible compression of its information content. Moreover, it is impossible in principle to prove that a given sequence is random, although it is clearly possible to demonstrate that it is non-random simply by finding a compression. Thus it will never be possible for us to prove that the sum total of information contained in all the laws of Nature might not be expressible in some more succinct form, which we shall refer to as the 'Secret of the Universe'. Of course, there may exist no such secret, and even if there does its information content may be buried very deep so that it takes a vast (or even infinite) amount of time to extract useful information from it by computation.

The question of the existence of a 'Secret of the Universe' amounts to discovering whether there is some deep principle from which all other knowledge of the physical world follows. A slightly weaker 'secret' would be the single proposition from which the largest amount of information follows. It is interesting to speculate as to the possible form of such a proposition. Would it be what philosophers call an 'analytic' statement or a 'synthetic' one? An analytic statement requires us to analyse the statement alone in order to ascertain its truth. An example is 'all bachelors are unmarried'. It is clearly a necessary truth, a consequence of logic alone. Synthetic statements are meaningful statements which are not analytic. The physical theories that we employ to understand the Universe are always synthetic. They tell us things that can only be checked by looking at the world. They are not logically necessary. They assert something about the world, whereas analytic statements do not. Some seekers after the Theory of Everything would seem to be hoping that the uniqueness and completeness of some particular mathematical theory will make it the only logically consistent description of the world and this will transform it from being a synthetic to an analytic statement. However, if we want the 'Secret of the Universe' to have testable predictions, it must be a synthetic statement. Yet this is not an entirely satisfactory conclusion because our 'secret' must then contain some ingredients that need to be deduced from some more fundamental principle, and so it cannot be *the* secret of the structure of the entire Universe: for it possesses ingredients that require further explanation by some deeper principle.

This dilemma extends to the problem of the role of mathematics in physics. If all mathematical statements are analytic—tautological consequences of some set of rules and axioms—then we are faced with trying to obtain synthetic statements about the world from purely analytic mathematical statements. In practice, if initial conditions remain unspecified by some form of self-consistency, then they supply a synthetic element that must be added to any analytic mathematical structure defined by differential equations. Even schemes like the 'no-boundary' condition, outlined in Chapter 3, simply introduce certain new 'laws' of physics as axioms.

What is it that makes necessary truths necessary? Presumably it is the feature that they are knowable *a priori*. If we have to carry out some act of observation to see whether a statement is true, then we can only know its truth *a posteriori*. A famous philosophical issue is that of whether all *a priori* statements are analytic. Most of the statements that we encounter in life are either synthetic *a posteriori* or analytic *a priori*. But are there non-analytic statements about the world which have real information content and which are knowable *a priori*? Is a synthetic *a priori* really possible? The most awkward problem would now appear to be how we could know that such a statement was giving us non-trivial information about the world without making some new observation to check. Traditionally, philosophical empiricists have maintained that synthetic *a priori* truths cannot exist, whilst rationalists have maintained that they do, although they have not been able to agree as to what they are. Ever since Immanuel Kant introduced this distinction between analytic and synthetic statements, there have been candidates for a synthetic *a priori* that have since been dispatched to oblivion, statements like 'parallel lines never meet' or 'every event has a cause', which were proposed before the advent of non-Euclidean geometry and the quantum theory.

How then can we have some form of synthetic *a priori* knowledge about the Universe? Kant suggested that the human mind is constructed in such a way that it naturally grasps some synthetic *a priori* aspects of the world. Whereas the real world possesses unimaginable features, our minds naturally sift out certain aspects of reality as though we were wearing rose-coloured spectacles. Our minds will only capture certain aspects of the world and this knowledge is thus synthetic and *a priori*. For it is an *a priori* truth that we will never understand anything that does not register in our particular mental categories. Hence, for us, there are certain necessary truths about the observable world. We might hope to flesh out this type of idea in a different way by considering the fact that there have been found to exist necessary cosmological conditions for the existence of observers in the Universe. These 'anthropic' conditions which we introduced earlier point us toward certain properties that the Universe must possess *a priori*, but which are non-trivial enough to be counted as synthetic. The synthetic *a priori* begins to look like the requirement that every knowable physical principle that forms part of the 'Secret of the Universe' must not forbid the possibility of our knowing it. The Universe is a member of the collection of mathematical concepts; but only those concepts with complexity sufficient to contain sub-programs which can represent 'observers' will be actualized in physical reality.

IS THE UNIVERSE A COMPUTER?

Mathematics is the part of science you could continue to do if you woke up tomorrow and discovered the universe was gone.
— DAVE RUSIN

There are two great streams of thought in contemporary science that, after running in parallel for so long, have begun to follow tantalizing channels that point to their future convergence. The circumstances of this coming together will determine which of them will ever after be seen as a mere tributary of the other. On the one side is the physicist's belief in the 'laws of Nature', guilded with symmetry, as the most fundamental bedrock of logic in the Universe. These symmetries are wedded to the picture of space and time as indivisible continua. Set against this is some picture of abstract computation, rather than symmetry, as the most fundamental of all notions. This image of reality portrays the logic at its bedrock as governing something *discrete* rather than continuous. The great unsolved puzzle for the future is to decide which is more fundamental: symmetry or computation. Is the Universe a cosmic kaleidoscope or a cosmic computer, a pattern or a program? Or neither? The choice requires us to know whether the laws of physics constrain the ultimate capability of abstract computation. Do they limit its speed and scope? Or do the rules governing the process of computation control what laws of Nature are possible?

Before discussing what little we can say about this choice it is good to be on one's guard concerning the choice itself. Throughout the history of human thought, there have been dominant paradigms of the Universe. These mental images often tell us little about the Universe, but much about the society that was engaged in its study. For those early Greeks who had developed a teleological perspective on the world as a result of the first systematic studies of living things, the world was a great organism. To others, who held geometry to be revered above all other categories of thought, the Universe was a geometrical harmony of perfect shapes. Later, in the era when the first clockwork and pendulum mechanisms were made, the image of the post-Newtonian Universe as a mechanism held sway, and launched a thousand ships of religious apologetic in search of the cosmic Clockmaker. For the Victorians of the industrial revolution, the prevailing paradigm was that of the heat engine and the physical and philosophical questions it raised concerning the laws of thermodynamics and the ultimate fate of the Universe bear the stamp of that age of machines. So, today, perhaps the image of the Universe as a computer is just the latest predictable extension of our habits of thought. Tomorrow, there may be a new paradigm. What will it be? Is there some deep and simple concept that stands behind logic in the same manner that logic stands behind mathematics and computation?

At first, the notions of symmetry and computation seem far removed and the choice between them appears a stark one. But symmetries dictate the possible changes that can occur and the 'laws' that result might be viewed as a form of software which run on some hardware, the material 'hardware' of our physical universe. Such a view implicitly subscribes to one of the particular views of the relationship between the laws of Nature and the physical universe which we introduced in Chapter 2. It regards the two as disjoint, independent conceptions. Thus, one could envisage the software being run upon different hardware. This view then seems to lead us into potential conflict with a belief in some unique Theory of Everything which unites the conditions for the existence of elementary particles to the laws that govern them.

The success of the continuum view of the physicist in explaining the physical world appears at first sight to argue against the discrete computational perspective. But logicians have waged a war of attrition against the notion of the number continuum during the last fifty years. Logicians like Quine claim that

> just as the introduction of the irrational numbers...is a convenient myth [which] simplifies the laws of arithmetic...so physical objects, are postulated entities which round out and simplify our account of the flux of existence...The conceptual scheme of physical objects is a convenient myth, simpler than the literal truth and yet containing that literal truth as a scattered part.

As yet, we have not found the right question to ask of the Universe whose answer will tell us whether computation is more primitive than symmetry: whether, in John Wheeler's words we can get

IT from BIT.

My personal view is that this hope cannot be completely fulfilled. In order for it to be found that computation is the most basic aspect of reality, we would require that the Universe only do computable things. The scope of the Universe's mathematical manifestations would be constrained to lie within the remit of the constructivists. This is the penalty of giving up the continuum and appealing to computable aspects of the world as the basis for explaining the whole. Yet we have uncovered many non-computable mathematical operations and physicists have found many to lurk within that piece of mathematics that is currently required for our understanding of the physical world. In the study of quantum cosmology, there have been found examples where a quantity which is observable in principle is predicted to have a value which equals an infinite sum of variable quantities each of which is to be evaluated on a particular type of surface. However, the listing of the required surfaces is proved to be a non-computable operation. It cannot be produced systematically by a finite number of computational steps of the Turing type. An element of novelty is required to generate each member of the set. There may, of course, be another way to calculate the observable quantity in question which bypasses the need to carry out this non-computable operation, but maybe not. Then there are further features of the discontinuous world in which discrete computation lives that actually make computability less likely.

Suppose that we take a simple ordinary differential equation of the sort

$$\mathrm{d}y/\mathrm{d}x = F(x, y),$$

that lies at the heart of all physical theories, where $F(x, y)$ is a continuous function of x and y which cannot be differentiated twice. This means that, although we can draw the curve F without taking our pencil point off the paper (the property of continuity), it can possess creases and sharp corners like we find at the point of a cone. Then, even though F is a computable function itself, there need exist no computable solution to the differential equation. If we examine partial differential equations which describe the propagation of waves of any sort, whether they be quantum waves or gravitational waves rippling through the geometry of space-time, then we encounter the same problem. When the initial wave profile is described by a continuous but not twice differentiable function, then there may exist no computable solution of the wave equation in two or more space dimensions. It is the lack of smoothness in the initial profile that is the crux of the problem. If the initial profile is twice differentiable, then all the solutions of the wave equations are computable. But, if at the most fundamental level, things are discrete and discontinuous, then we may fall foul of the problem of non-computability.

The answers to these difficulties, if they can be found, surely lie in an enlarged concept of what we mean by a computation. It turns out, for example, that there is a way of characterizing all badly behaved differential equations like the one above so that they are computable on the average even though there may be a worse-case scenario where they are not. Traditionally, computer scientists have defined the ultimate capability of any computer, whether real or imaginary, to be that of Turing's idealized machine. Indeed, the capability of such a machine defines what we mean by the accolade 'computable'. Yet in recent years it has become clear that one can fabricate computers which are intrinsically quantum mechanical in nature and so exploit the quantum uncertainties of the world to perform operations which are beyond the capability of Turing's idealized machine. Since the world is at root a quantum system, any attempt to explain its inner workings in terms of the computational paradigm must be founded upon a firm understanding of what quantum computation actually is and what it can achieve that a conventional Turing machine cannot. In many ways, the computational paradigm has an affinity for the quantum picture of the world. Both are discrete; both possess dual aspects like evolution and measurement (compute and read). But greater claims could be made for the relationship between the quantum and the symmetries of Nature. Half a century of detailed study by physicists has wedded the two into an indissoluble union. What might be the status of the computational paradigm after a similar investment of thought and energy?

THE UNKNOWABLE

I hate quotations. Tell me what you know.
— RALPH WALDO EMERSON

'Why is the world mathematical?' we ask. But, on second thoughts, don't many of the things we encounter in everyday life seem more like almost anything but mathematics. Mathematics is relegated to the description of a peculiar skeletal world that we are assured lies behind the mere appearances, a world that is simpler than the one in which we are daily participants. Yet we find nothing mathematical about emotions and judgements, about music or art. How then, when we speak of 'Theories of Everything' and pursue them with mathematics confident that all diversity will evaporate to leave nothing but number, can we draw the line that divides those elusive phenomena which are intrinsically non-mathematical from those which are encompassed by a Theory of Everything? What are the things that cannot be included in the physicist's conception of 'everything'? There appear to be such things, but they are more often than not excluded from the discussion on the grounds that they are not 'scientific'—a response not unlike that of the infamous Master of Balliol of whom it was said that 'what he doesn't know isn't knowledge'.

We all have a good idea of the direction to which we should look in order to outflank a Theory of Everything. The very response of our minds to being dealt certain varieties of information provides a suggestive clue. The late Heinz Pagels has written of his differing experiences when reading 'factual' scientific writing as opposed to the subjective comment one might find in the literary pages of the newspapers:

> Once I was at a New York City dinner party with a group of well-educated people. They were writers, editors and intellectuals; not a scientist in the group except for me. Somehow the conversation got around to *The New York Review of Books*, a fine book review magazine that went well beyond just reviewing books . . . I avidly read and liked it . . . But I went on to describe my problem: I couldn't remember anything that I read in it. The information went into short-term memory storage and never got into my long-term memory. The reason for this, I had decided, was that in spite of the consistently brilliant style of writing, and the quality of the

narrative, all that was being expressed in effect was one person's opinion about another person's thinking or actions. It is difficult for me to remember people's opinions (even my own). What I remember are concepts and facts, the invariants of experience, not the ephemera of human opinion, taste, and styles. Such trivia are not to be considered by serious people, except as intellectual recreation.

Silence followed my brief remarks, and I felt isolated. The rift between the two cultures—science and humanism—widened considerably. I realized that in my blundering I had violated the sacred precincts of the other guests' high temple. Those people worshipped in that temple which was dedicated to political opinion, taste, and style, to a consciousness dominated by self-reflection, belief and feeling, and intellectual gossip and activity for its own sake, only loosely bound by the constraints of knowledge. I tried to think of a joke to extract myself from an awkward situation but could not.

What this rather telling piece of introspection reveals is that Pagels detects a personal difficulty in extracting and organizing the content of some varieties of information. As a scientist, his mind has been trained to act and respond in certain ways to particular types of input. Whereas factual or logically structured information has a ready-made framework within which it can be accommodated, other sorts of information do not. They defy compression into ordered and easily recallable forms. This tendency has suggestive facets.

We have already seen how the brain performs algorithmic compressions on the information made available to it. When strings of facts can be significantly algorithmically compressed, we are on the way to creating a 'science'. Clearly, some branches of experience are more amenable to this sublimation than others. In the 'hard' sciences, the most important feature of their subject matter that encourages algorithmic compression is the existence of sensible idealizations of complicated phenomena which underwrite very accurate approximations to the true state of affairs. If we wish to develop a detailed mathematical description of the observed features of a typical star like the Sun, then it is an excellent approximation to treat the Sun as being spherical with the same temperature all over its surface. Of course, no real star is *exactly* spherical and superficially isothermal in this way. But all stars are such that some collection of idealizations of this sort can be made and the resulting descriptions are very accurate. Subsequently, the idealizations can be relaxed a little and one can proceed towards a more realistic description that allows for the presence of small asphericities, then to one that introduces further realism, and so forth. Such a step-by-step sequence of better and better approximations to the phenomenon under study is what one means by a 'computable' operation in Turing's sense. By contrast, many of the 'soft' sciences which seek to apply mathematics to things like social behaviour, prison riots, or psychological reactions fail to produce a significant body of sure knowledge because their subject matters do not provide obvious and fruitful idealizations. Complicated phenomena, especially those which possess aspects that may be algorithmically incompressible, or which, like personal opinions, are intrinsically unpredictable because they react to being investigated, are not replaceable by simple approximations. It is not easy to see how one can model an 'approximate society' or 'approximate paranoia'. These phenomena do not permit the effective use of the mind's most successful device for making sense of complexity.

In practice, this may be a failing of our minds to find the right way to go about the search for idealizations or it may be a consequence of some intrinsic incompressibility associated with the phenomena in question. We know of many examples of the former situation of course; whenever we have a new idea that makes new sense of what was just a jumble of confusing facts, we witness the force of this possibility. Of the latter possibility, what can we say? Can we be sure that there are any examples in this category at all? What sort of things fail the test of mathematics?

Science is most at home attacking problems that require technique rather than insight. By technique we mean the systematic application of a sequential procedure—a recipe. The fact that this approach to the world so often bears fruit witnesses to the power of generalization. Nature uses the same basic idea again and again in different situations. The hallmark of these reapplications is their mathematical character. The search for the Theory of Everything is a quest for that technique whose application could decode the message of Nature in every circumstance. But we know there must exist circumstances where mere technique will fail.

The American logician John Myhill has proposed a metaphorical extension of the lessons that we have learnt from the theorems of Gödel, Church, and Turing about the scope and limitations of logical systems. The most accessible and quantifiable aspects of the world have the property of being computable. There exists a definite procedure for deciding if any given candidate either does or does not possess the required property. Human beings can be trained to respond to the presence or absence of this sort of property. Truth is not in general such a property of things; being a prime number is. A more elusive set of properties are those that are merely listable. For these, we can construct a procedure which will list all the quantities which possess the required property (even though you might have to wait an infinite time for the listing to end), but there is no way of systematically generating all the entities which do not possess the required property. Most logical systems have the property of being listable but not computable: all their theorems can be listed but there is no automatic procedure for inspecting a statement and deciding whether it is or is not a theorem. If the mathematical world had no Gödel theorem then every property of any system that contained arithmetic would be listable. We could write a definite program to carry out every activity. Without the restrictions of Turing and Church on computability, every property of the world would be computable. The problem of deciding whether this page is an example of grammatical English is a computable one. The words can be checked against a reference dictionary and the grammatical constructions employed could be checked sequentially. But the page of text could still be meaningless to a reader who did not know English. As time passes, this reader could learn more of the English language and more and more of the page would become meaningful to him. But there is no way of predicting ahead of time which bits of this page they will be. The property of meaningfulness is thus listable but not computable. Likewise, the question of whether this page is something that the reader might want to write in the future is also a listable but not a computable property.

Not every feature of the world is either listable or computable. For example, the property of being a true statement in a particular mathematical system is neither listable nor computable. One can approximate the truth to greater and greater accuracy by introducing more and more rules of reasoning and adding further axiomatic assumptions, but it can never be captured by any finite set of rules. These attributes that have neither the property of listability nor that of computability—the 'prospective' features of the world—are those which we cannot recognize or generate by a series of sequence of logical steps. They witness to the need for ingenuity and novelty; for they cannot be encompassed by any finite collection of rules or laws. Myhill reminds us that Beauty, simplicity, truth, these are all properties that are prospective. There is no magic formula that can be called upon to generate all the possible varieties of these attributes. They are never fully exhaustible. No program or equation can generate all beauty or all ugliness; indeed, there is no sure way of recognizing either of these attributes when you see them. The restrictions of mathematics and logic prevent these prospective properties falling victim to mere technique even though we can habitually entertain notions of beauty or ugliness. The prospective properties of things cannot be trammelled up within any logical Theory of Everything. No non-poetic account of reality can be complete.

The scope of Theories of Everything is infinite but bounded; they are necessary parts of a full understanding of things but they are far from sufficient to reveal everything about a Universe like ours. In the pages of this book, we have seen something of what a Theory of Everything might hope to teach us about the unity of the Universe and the way in which it may contain elements that transcend our present compartmentalized view of Nature's ingredients. But we have also learnt that there is more to Everything than meets the eye. Unlike many others that we can imagine, our world contains prospective elements. Theories of Everything can make no impression upon predicting these prospective attributes of reality; yet, strangely, many of these qualities will themselves be employed in the human selection and approval of an aesthetically acceptable Theory of Everything.

There is no formula that can deliver all truth, all harmony, all simplicity. No Theory of Everything can ever provide total insight. For, to see through everything, would leave us seeing nothing at all.

Select Bibliography

The following bibliography is designed to allow the interested reader to delve further into the topics covered in the earlier chapters of the book. The works cited are at variety of different levels: some are designed for the general reader, whilst others are suitable only for those who possess some background in mathematics and physics.

CHAPTER I

Barrow, J. D., 'Inner Space and Outer Space', in *The Centenary Gifford Lectures*, ed. N. Spurway (Blackwell, Oxford, 1991).

Bettelheim, B., *The Uses of Enchantment* (Penguin, Harmondsworth, 1978).

Blacker, C. and Loewe, M. (eds.), *Ancient Cosmologies* (Allen & Unwin, London, 1975).

Chaitin, G., *Algorithmic Information Theory* (CUP, Cambridge, 1987).

De Santillana, G. and Von Dechend, H., *Hamlet's Mill: An Essay on Myth and the Frame of Time* (Macmillan, London, 1969).

Dobzhansky, T., *The Biology of Ultimate Concern* (Meridian, New York, 1967).

Frazer, J., *The Golden Bough* (Macmillan, New York, 1922).

La Rue, G. A., *Ancient Myth and Modern Man* (Prentice-Hall, NJ, 1975).

Laurikainen, K. V., *Beyond the Atom: the Philosophical Thought of Wolfgang Pauli* (Springer, Berlin, 1988).

Lewis, C. S., *The Discarded Image* (CUP, Cambridge, 1964).

Lloyd, S. and Pagels, H., 'Complexity as Thermodynamic Depth', *Annals of Physics* (*New York*) **188**, 186 (1988).

McAllister, J. W., 'Truth and Beauty in Scientific Reason', *Synthèse* **78**, 25 (1989).

Maclagen, D., *Creation Myths* (Thames & Hudson, London, 1977).

Pagels, H., *The Cosmic Code* (Schuster, New York, 1982).

Pagels, H., *The Dreams of Reason* (Simon & Schuster, New York, 1989).

Piaget, J., *Child's Conception of Number* (Humanities Press, NJ, 1964).

Rescher, N., 'Some Issues Regarding the Completeness of Science and the Limits of Scientific Knowledge', in *The Structure and Development of Science*, eds. G. Radnitzky and G. Anderson (Reidel, Dordrecht, 1979), pp. 15–40.

Smith, J. W., *Essays on Ultimate Questions* (Avebury, Aldershot, 1988).

Turbayne, C. M., *The Myth of Metaphor* (Yale University Press, New Haven, 1962).

von Weizsäcker, C. F., *The Relevance of Science* (Collins, London, 1964).

Yates, F., *Giordano Bruno and the Hermetic Tradition* (Routledge & Kegan Paul, London, 1964).

CHAPTER 2

Barrow, J. D., *The World within the World* (OUP, Oxford, 1988).

Barrow, J. D., *The Infinite Book* (Jonathan Cape, London, 2005).

Boscovich, R. J., *A Theory of Natural Philosophy*, English translation of Venice edition of 1763 (MIT Press, Cambridge MA, 1966).

Chaitin, G., 'Randomness in Arithmetic', *Scientific American*, July 1988, p. 80.

Cobb, J. B. and Griffin, D. R., *Process Theology: An Introductory Exposition* (Westminster Press, London, 1976).

Feynman, R., *The Character of Physical Law* (MIT Press, Cambridge MA, 1965).

Funkenstein, A., *Theology and the Scientific Imagination from the Middle Ages to the Seventeenth Century* (Princeton University Press, Princeton, 1986).

Greene, B., *The Elegant Universe: Superstrings, Hidden Dimensions, and the Quest for the Ultimate Theory* (Vintage, NY, 2000).

Jaki, S., *The Relevance of Physics* (University of Chicago Press, Chicago, 1966).

Needham, J., *The Grand Titration: Science and Society in East and West* (Allen & Unwin, London, 1969).

Oerter, R., *The Theory of Almost Everything: The Standard Model, the Unsung Triumph of Modern Physics* (Pi Books, London, 2005).

Stenger, V., *The Comprehensible Universe: Where do the Laws of Physics Come From?* (Prometheus, NY, 2006).

CHAPTER 3

Barrow, J. D. and Silk, J., *The Left-Hand of Creation* (Basic Books, New York, 1983).

Bondi, H., *Cosmology* (CUP, Cambridge, 1953).

Bondi, H., 'The Steady-State Theory of the Universe', in *Rival Theories of Cosmology*, eds. H. Bondi, W. B. Bonnor, R. A. Lyttleton, and G. J. Whitrow (OUP, Oxford, 1960).

Chaitin, G., *MetaMath: The Quest for Omega* (Pantheon, NY, 2005).

Craig, W. L., *The Cosmological Argument from Plato to Leibniz* (Macmillan, London, 1980).

Davidson, H. A., *Proofs for Eternity, Creation and the Existence of God in Medieval Islamic and Jewish Philosophy* (OUP, New York, 1987).

Drees, W., *Beyond the Big Bang: Quantum Cosmology and God*, Ph.D. thesis, University of Groningen (Open Court, La Salle, 1990).

Franzen, T., *Gödel's Theorem: An Incomplete Guide to Its Use and Abuse* (A. K. Peters, NY, 2005).

Grünbaum, A., 'The Pseudo-Problem of Creation in Cosmology', *Philosophy of Science* **56**, 373 (1989).

Guth, A., 'The Inflationary Universe: A Possible Solution to the Horizon and Flatness Problems', *Physical Review* D 23, 347 (1981).

Guth, A., *The Inflationary Universe* (Addison-Wesley, Reading, 1997).

Guth, A. and Steinhardt, P., 'The Inflationary Universe', *Scientific American*, May 1984, p. 116.

Hartle, J. B. and Hawking, S. W., 'Wave Function of the Universe', *Physical Review* D 28, 2960 (1983).

Hawking, S. W., 'The Quantum Theory of the Universe', in *Interactions between Elementary Particle Physics and Cosmology*, eds. T. Piran and S. Weinberg (World Scientific Press, Singapore, 1986).

Hawking, S. W., *A Brief History of Time* (Bantam, New York, 1988).

Kaku, M., *Parallel Worlds* (Penguin, London, 2005).

McCrea, W. H., 'The Interpretation of Cosmology', *La Nuova Critica (III Serie) Quaderno* XI, 11 (1960).

Nasr, S., *Introduction to Islamic Cosmological Doctrines* (Harvard University Press, Cambridge MA, 1964).

Philo, *On the Account of the World's Creation Given By Moses*, translated by F. H. Colson and G. H. Whitaker (Heinemann, London, 1981).

Polkinghorne, J., *Science and Creation; the Search for Understanding* (SPCK, London, 1988).

Raatilainen, P., 'Algorithmic Information Theory and Undecidability', *Synthèse* 123, 217–25 (2000).

Russell, R. J., Stoeger, W. R., and Coyne, G. V., *Physics, Philosophy and Theology* (University of Notre Dame Press, IN, 1988).

Smolin, L., *The Trouble With Physics: The Rise of String Theory, the Fall of a Science, and What Comes Next* (Houghton Mifflin Company, NY, 2006).

Sorabji, R., *Time, Creation and the Continuum* (Duckworth, London, 1983).

Tipler, F. J., 'The Omega Point as Eschaton', *Zygon* 24, 217 (1989).

Vilenkin, A., 'Creation of Universes from Nothing', *Physics Letters* B 117, 25 (1982).

Vilenkin, A., 'Boundary Conditions in Quantum Cosmology', *Physical Review* D 33, 3560 (1982).

Whitehead, A. N., *Adventures of Ideas* (CUP, New York, 1933).

Whitehead, A. N., *Science and the Modern World* (CUP, London, 1953).

Woit, P., *Not Even Wrong: The Failure of String Theory and the Continuing Challenge to Unify the Laws of Physics* (Jonathan Cape, London, 2006).

CHAPTER 4

Atiyah, M., 'Geometry, Topology and Physics', *Quarterly Journal of the Royal Astronomical Society* 29, 287.

Bailin, D., 'Why Superstrings', *Contemporary Physics* 30, 237 (1989).

Barrow, J. D., *The Book of Nothing* (Jonathan Cape, London, 2000).

Cooper, N. C. and West, G., *Particle Physics: A Los Alamos Primer* (CUP, Cambridge, 1988).

Green, M., 'Superstrings', *Scientific American*, September 1986, p. 48.

Greene, B., *The Fabric of the Cosmos: Space, Time, and the Texture of Reality* (Knopf, NY, 2005).

Harman, P. M., *Energy, Force and Matter: The Conceptual Development of Nineteenth Century Physics* (CUP, Cambridge, 1982).

Jammer, M., *Concepts of Force* (Harvard University Press, Cambridge MA, 1957).

Krauss, L., *Hiding in the Mirror: The Mysterious Allure of Extra Dimensions, from Plato to String Theory and Beyond* (Viking, NY, 2005).

Pagels, H., *Perfect Symmetry* (Michael Joseph, London, 1985).

Peat, F. D., *Superstrings and the Search for the Theory of Everything* (Contemporary Books, Chicago, 1988).

Schwartz, J. H., 'Superstring Unification', in *300 Years of Gravitation*, eds. S. W. Hawking and W. Israel (CUP, Cambridge, 1987), p. 652.

Susskind, L., *The Cosmic Landscape: String Theory and the Illusion of Intelligent Design* (Little Brown, NY, 2005).

Thomson, W., 'On Vortex Atoms', *Philosophical Magazine* 34, 15 (1867).

Wilczek, F., and Devine, B., *Longing For The Harmonies* (Norton, New York, 1988).

Wilczek, F., 'Gauge Theories of Swimming', *Physics World* 2, 36 (1989).

Zee, A., *Fearful Symmetry: The Search for Beauty in Modern Physics* (Macmillan, New York, 1986).

CHAPTER 5

Barrow, J. D., 'Observational Limits on the Time-Evolution of Extra Spatial Dimensions', *Physical Review* D 35, 1805 (1987).

Barrow, J. D., 'Constants of Physics and the Structure of the Universe', in *Saas Fee Lectures on Unités de Mesure et Constants Physique*, eds. M. Batato, R. Behn, J-F. Loude, and H. Weisen (Lausanne, Association Vaudoise des Chercheurs en Physique), chap. 5.

Barrow, J. D., 'The Mysterious Lore of Large Numbers', in *Modern Cosmology in Retrospect*, eds. S. Bergia and B. Bertotti (CUP, Cambridge, 1990).

Barrow, J. D., *The Constants of Nature* (Jonathan Cape, London, 2002).

Barrow, J. D. and Tipler, F. J., *The Anthropic Cosmological Principle* (OUP, Oxford, 1986).

Barrow, J. D. and Webb, J. K., 'Inconstant Constants', *Scientific American*, June 2005, pp. 56–63.

Carr, B. J. and Rees, M. J., 'The Anthropic Principle and the Structure of the Physical World', *Nature* 278, 605 (1979).

Coleman, S., 'Black Holes as Red Herrings: Topological Fluctuations and the Loss of Quantum Coherence', *Nuclear Physics* B 307, 867 (1988).

Coleman, S., 'Why There is Nothing rather than Something: A Theory of the Cosmo-logical Constant', *Nuclear Physics* **B 310**, 643 (1988).

Douglas, A. V., *The Life of Arthur Stanley Eddington* (Nelson, London, 1956).

Eddington, A. S., *New Pathways in Science* (CUP, Cambridge, 1935).

Eddington, A. S., *Fundamental Theory* (CUP, London, 1946).

Einstein, A., 'Physics and Reality', *Journal of the Franklin Institute* **221**, 349 (1936).

Hawking, S. W., 'Wormholes in Space-time', *Physical Review* **D 37**, 904 (1988).

Hawking, S. W., 'Baby Universes', *Modern Physics Letters* **A 5**, 453 (1990).

Jungnickel, C. and McCormmach, R., *Intellectual Mastery of Nature: Theoretical Physics from Ohm to Einstein*, vols. 1 and 2 (University of Chicago Press, Chicago, 1986).

Levy-Leblond, J. M., 'Constants of Physics', *Rivista Nuovo Cimento* **7**, 187 (1977).

McCrea, W. H. and Rees, M. J. (eds.), *The Constants of Physics* (The Royal Society, London, 1983).

Pais, A., *Subtle is the Lord: The Science and Life of Albert Einstein* (OUP, Oxford, 1982).

Rosenthal-Schneider, I., *Reality and Scientific Truth: Discussions with Einstein, Von Laue and Planck* (Wayne State Press, PA, 1980).

Weinberg, S., 'The Cosmological Constant Problem', *Reviews of Modern Physics* **61**, 1 (1989).

Witt-Hansen, J., *Exposition and Critique of the Concepts of Eddington Concerning the Philosophy of Physical Science* (Gads, Copenhagen, 1958).

Yolton, J., *The Philosophy of Science of A. S. Eddington* (Nijhoff, The Hague, 1960).

CHAPTER 6

Bartholomew, D. J., *God of Chance* (SCM, London, 1984).

Bartholomew, D. J., 'Probability, Statistics and Theology', *Journal of the Royal Statistical Society* **A 151**, 137 (1988).

Ford, J., 'How Random is a Coin Toss?', *Physics Today*, April 1983, p. 40.

Campbell, L. and Garnett, W., *The Life of James Clerk Maxwell* (London, 1882; reprinted by the Johnson Reprint Corporation, New York, 1969).

Gleick, J., *Chaos: Making A New Science* (Viking, New York, 1987).

Hacking, I., *The Emergence of Probability* (CUP, Cambridge, 1975).

Laughlin, R., *A Different Universe: Reinventing Physics from the Bottom Down* (Basic Books, NY, 2006).

Linde, A., 'Particle Physics and Inflationary Cosmology', (Gordon & Breach, New York, 1989).

Pearson, K., *The History of Statistics in the Seventeenth and Eighteenth Centuries, against the Changing Background of Intellectual, Scientific and Religious Thought*, ed. E. S. Pearson (Griffin, London, 1978).

Sambursky, S., 'On the Possible and Probable in Ancient Greece', *Osiris* **12**, 35 (1956).

Sheynin, O. B., 'On the Prehistory of the Theory of Probability', *Archive for the History of the Exact Sciences* **12**, 97 (1974).

Stewart, I., *Does God Play Dice: The Mathematics of Chaos* (Blackwell, Oxford, 1989).

Stewart, I., *What Shape is a Snowflake?* (Weidenfeld & Nicholson, London, 2001).

CHAPTER 7

Ayala, F. J. and Dobzhansky, T. (eds.), *Studies in the Philosophy of Biology* (Macmillan, London, 1974).

Bak, P., *How Nature Works: The Science of Self-organised Criticality* (OUP, Oxford, 1996)

Ball, P., *The Self-made Tapestry* (OUP, Oxford, 2001).

Ball, P., *Critical Mass: How One Thing Leads to Another* (Arrow, London, 2005).

Bohm, D., *Wholeness and the Implicate Order* (Routledge & Kegan Paul, London, 1980).

Clayton, P., *Mind and Emergence: From Quantum to Consciousness* (OUP, Oxford, 2006).

Clayton, P. and Davies, P. C. W., *The Re-emergence of Emergence: The Emergentist Hypothesis from Science to Religion* (OUP, Oxford, 2006).

Davies, P. C. W., *The Cosmic Blueprint* (Heinemann, London, 1987).

Delbrück, M., *Mind From Matter: An Essay on Evolutionary Epistemology* (Blackwell Scientific, Oxford, 1986).

Eigen, M. and Winkler, R., *Laws of the Game: How the Principles of Nature Govern Chance* (Penguin, Harmondsworth, 1983).

Gell-Mann, M., *The Quark and the Jaguar* (W. H. Freeman, NY, 1994).

Kauffman, S., *At Home in The Universe: The Search for the Laws of Self-organisation and Complexity* (OUP, NY, 1995).

Langton, C. G. (ed.), *Artificial Life* (Addison-Wesley, New York, 1989).

Leggett, A. J., *The Problems of Physics* (OUP, Oxford, 1987).

Minsky, M., *The Society of Mind* (Simon & Schuster, New York, 1986).

Moravec, H., *Mind Children* (Harvard University Press, Cambridge MA, 1988).

Morowitz, H., *The Emergence of Everything: How the World Became Complex* (OUP, NY, 2002).

Zeh, H., *The Physical Basis of the Direction of Time* (Springer, Berlin, 1989).

CHAPTER 8

Barrow, J. D., 'Life, the Universe, and the Anthropic Principle, *World and I Magazine*, August 1987, p. 179.

Barrow, J. D., 'Patterns of Explanation in Cosmology', in *The Anthropic Principle*, eds. F. Bertola and U. Curi (CUP, Cambridge, 1989).

Barrow, J. D. and Tipler, F. J., *The Anthropic Cosmological Principle* (OUP, Oxford, 1986).

Carter, B., 'Large Number Coincidences and the Anthropic Principle in Cosmology', in *Confrontation of Cosmological Theories with Observational Data*, ed. M. Longair (Reidel, Dordrecht, 1974).

Carter, B., 'The Anthropic Principle: Self-selection as an Adjunct to Natural Selection', in *Cosmic Perspectives*, ed. C. V. Vishveshwara (CUP, Cambridge, 1989).

Carter, B., 'Anthropic Selection Principle and the Ultra-Darwinian Synthesis' in *The Anthropic Principle*, eds. F. Bertola and U. Curi (CUP, Cambridge, 1989).

Davies, P. C. W., *The Physics of Time Asymmetry* (University of California Press, Berkeley, 1974).

Davies, P. C. W., *The Goldilocks Enigma* (Penguin, London, 2006).

Ellis, G. F. R. (ed.), *The Far-Future Universe* (Templeton Press, Radnor, 2002).

Leslie, J., 'Observership in Cosmology: The Anthropic Principle', *Mind* 92, 573, 1983.

Leslie, J., *Universes* (Macmillan, London, 1989).

Linde, A., 'The Universe: Inflation Out of Chaos', *New Scientist*, March 1985, p. 14.

Nicolis, G. and Prigogine, I., *Self-organization in Non-equilibrium Systems* (Wiley, New York, 1977).

Page, D. N., 'The Importance of the Anthropic Principle', *World and I Mazazine*, August 1987, p. 392.

Piaget, J., *Genetic Epistemology* (Norton, New York, 1971).

Polanyi, M., 'Life's Irreducible Structure', *Science*, June 1968, p. 1308.

Prigogine, I., *From Being to Becoming: Time and Complexity in the Physical Sciences* (Freeman, San Francisco, 1980).

Prigogine, I. and Stengers, I., *Order Out of Chaos* (Heinemann, London, 1984).

Rees, M. J., *Before the Beginning* (Simon & Schuster, NY, 1997).

Smolin, L., *The Life of the Cosmos* (OUP, Oxford, 1997).

Smolin, L., 'Did the Universe Evolve?', *Class. Quantum Gravity* 9, 173 (1984).

Sayers, D., *Have His Carcase* (Victor Gollancz, London, 1932).

Vollmer, G., 'Mesocosm and Objective Knowledge', in *Concepts and Approaches in Evolutionary Epistemology*, ed. F. W. Wuketits (Reidel, Dordrecht, 1984).

Stent, G. S., 'Light and Life: Niels Bohr's Legacy to Contemporary Biology', in *Niels Bohr: Physics and the World*, eds. H. Feshbach, T. Matsui, and A. Oleson (Harwood Academic, New York, 1988).

Weinberg, S. W., 'Newtonianism, Reductionism and the Art of Congressional Testimony', *Nature* 330, 433 (1987).

CHAPTER 9

Barrow, J. D., 'The Mathematical Universe', *World and I Magazine*, May 1989, p. 306.

Bennett, C. H., 'The Thermodynamics of Computation—A Review', *International Journal of Theoretical Physics* 21, 905 (1982).

Bennett, C. H. and Landauer, R., 'The Fundamental Physical Limits of Computation', *Scientific American* 253, pp. 48, 253 (issue 1) and p. 6 (issue 4) (1985).

Birkhoff, G., 'The Mathematical Nature of Physical Theories', *American Scientist* 31, 281 (1943).

Casti, J. and Karlqvist, A. (eds.), *Boundaries and Barriers: On the Limits to Scientific Knowledge* (Addison-Wesley, Reading, 1996).

Davies, P. C. W., 'Why is the Universe Knowable?' in *Maths and Science*, ed. R. E. Mickens (OUP, New York, 1989).

Davis, P. and Hersch, R., *Descartes' Dream* (Penguin, Harmondsworth, 1989).

Deutsch, D., 'Quantum Theory, the Church–Turing Principle, and the Universal Quantum Computer', *Proceedings of the Royal Society of London* A 400, 97 (1985).

Deutsch, D., 'On Wheeler's notion of "Law Without Law" in Physics', in *Between Quantum and Cosmos*, eds. W. Zurek, A. van der Merwe, and W. A. Miller (Princeton University Press, Princeton, 1988), pp. 583–92.

Deutsch, D., 'Quantum Communication Thwarts Eavesdroppers', *New Scientist*, 9 December 1989, p. 25.

Dewdney, A. K., *The Turing Omnibus* (Computer Science Press, Rockville, 1989).

Dyson, F., 'Mathematics in the Physical Sciences', *Scientific American*, September 1964, p. 129.

Field, H., *Science without Numbers* (Blackwell, Oxford, 1980).

Field, H., *Realism, Mathematics and Modality* (Blackwell, Oxford, 1989).

Hadamard, J., *The Psychology of Invention in the Mathematical Field* (Princeton University Press, Princeton, 1945).

Hersh, R., *What is Mathematics, Really?* (Cape, London, 1997).

Kitcher, P., *The Nature of Mathematical Knowledge* (OUP, New York, 1983).

Kline, M., *Mathematics and the Search for Knowledge* (OUP, New York, 1985).

Kramer, E. E., *The Nature and the Growth of Modern Mathematics* (Princeton University Press, Princeton, 1982).

Landauer, R., 'Dissipation and Noise Immunity in Computation and Communication', *Nature* 335, 779 (1988).

Lehman, H., *Introduction to the Philosophy of Mathematics* (Blackwell, Oxford, 1975).

Lloyd, S., *Programming the Universe* (Jonathan Cape, London, 2005).

Maddy, P., *Realism in Mathematics* (OUP, New York, 1992).

Myhill, J., 'Some Philosophical Implications of Mathematical Logic', *The Review of Metaphysics* 6, 165 (1952).

Penrose, R., *The Emperor's New Mind* (OUP, Oxford, 1989).

Penrose, R., *The Road to Reality* (Random House, London, 2004).

Rucker, R., *Mindtools* (Penguin, Harmondsworth, 1987).

Tipler, F. J., 'It's All in the Mind' (review of R. Penrose, *The Emperor's New Mind*), *Physics World*, November 1989, p. 45.

Traub, J. F. and Werschulz, A. G., 'Linear Ill-posed Problems are Solvable on the Average for All Gaussian Measures', *The Mathematical Intelligencer* 16 (2), 42–8 (1994).

Traub, J. F. and Woźniakowski, H., 'Breaking Intractability', *Scientific American*, January 1994, pp. 102–7.

Wigner, E., 'The Unreasonable Effectiveness of Mathematics in the Natural Sciences', *Communications in Pure and Applied Mathematics* 13, 1 (1960).

Wolfram, S., *A New Kind of Science* (Wolfram Inc., Champaign, 2002).

Wolfson, H. A., *Religious Philosophy* (Harvard University Press, Cambridge MA, 1961).

Zurek, W. H., 'Thermodynamic Cost of Computation, Algorithmic Complexity and the Information Metric', *Nature* 341, 119 (1989).

Index